Cry Me a River

Dedication

To my wife Carol; to Jonathan, Heidi and Amanda, thanks for putting up with me all these years. To the team who made the trip happen, I will be forever in your debt. A special thanks to Kate Stewart who made sure I put some emotion into the words, and to Lynne Ferguson who produced an effective proof read at lightning speed while convincing me the story was OK. Merri Mulherin from Newcastle came up with the name and Giovanni Ebono made the book happen. To Geoff McBride, who brought the trip to so many people on a daily basis with his tireless management of the web site. This book is dedicated to all of you.

Cry Me a River

**One man's journey down the
Murray Darling with a kayak on wheels**

Steve Posselt

Ebono Institute

First published in 2009 by the Ebono Institute,
17 Brunswick Terrace, Mullumbimby
Copyright © Kayak4earth Pty Ltd 2009

All Rights Reserved. No parts of this book may be reproduced or transmitted in any form or by any means, electronic or mechanical, including photocopying, recording or by information storage and retrieval system, without prior permission in writing from the publisher.

National Library of Australia, Cataloguing-in-publication data

Posselt, Steve (Stephen John)
    Cry Me a River: One man's journey down the Murray Darling
    with a kayak on wheels
    (pbk) Includes index

    ISBN 978 098061370 4

    1. Rivers – Australia
    2. Rivers – Environmental aspects – Australia
    3. Watersheds – Australia

333.91620994

Editor: Ebono, Giovanni

The names of some characters in this book have been changed. All quotes are printed verbatim and have been transcribed from recordings.

Ebono Institute

Designed and typeset in Berkeley 10/13 by Southern Star Design
Cover design by Graphiti Design, Lismore NSW
Printed in Australia by McPherson's Printing Group

Contents

Prologue 1

PART ONE 3
  1. An Idea Takes Shape 5
  2. Be Prepared, Be Very Prepared 19
  3. Brisbane – Ipswich 29
  4. Up the Hill to Toowoomba 41
  5. Toowoomba 55
  6. Onto the Darling Downs 65
  7. Daby – Chinchilla 75
  8. Condamine – Surat 87
  9. St George 109
  10. The Last of Queensland 125

PART TWO 147
  11. The Bokhara Delta 149
  12. Brewarrina – Bourke 159
  13. Bourke – Trevallyn 181
  14. Trevallyn – Wentworth 201

PART THREE 215
  15. Onto the Murray River 217
  16. Renmark – Waikerie 229
  17. The Last of the Murray 241
  18. The Southern Ocean – Adelaide 257

Epilogue 266
Index 268
About the Author 274
Sustaining Australia 277

Maps

Steve's journey through the Murray Darling Basin 4
1. Brisbane – Toowoomba 28
2. Toowoombah – St George 54
3. St George – Bourke 108
4. Bourke - Wentworth 180
5. Wentworth – Adelaide 216

Colour plates illustrating the text appear between pages 136 and 137

Foreword

Steve Posselt calls himself a civil engineer who, he says, happens to be able to read the writing on the wall, telling us that our rivers are dying. And after his extraordinary journey paddling and walking thousands of kilometres along the Murray Darling river system which he so entertainingly chronicles here, he speaks with authority. While his journey is exhilarating as he sweeps us along in his personable style, the way he describes the beauty of our landscape and its devastation, becomes a wake up call to everyone in Australia.

We simply cannot continue as we have been – burning fossil fuels, putting in infrastructure to sell our resources to other countries, having questionable irrigation practices and simply taking and taking without heeding the laws of nature. Steve explains how wetlands serve a purpose, how our rivers are the arteries of our landscape and how we must share water wisely – with each other, the wildlife and the landscape and that long term management of the natural systems is a necessary condition of our survival.

Steve's journey, his knowledge and experiences, are a beacon, a warning and, hopefully, the start of a solution. Join him on his travels through the pages of this book and learn, as I did, how close we are to midnight when our rivers will perish. Steve does not preach, but he is an acute and interesting observer who concludes …

> "We all want to build a way of life that benefits our children and our grandchildren. If what we build is not sustainable, then we have robbed them of their inheritance. From my observations that is exactly what we have done. Our river systems are precious. If they die, we die. And they *are* dying."

Thank you, Steve. I hear you and I cry too.

Di Morrissey, February 2009

What Others Say

This is a 'must read' for those interested in discovering how rivers really are the arteries of our country. Steve has chronicled his discovery of the current state of one of our greatest rivers and challenges us all to be a part of the remediation and protection of all rivers. It is a challenge that he has taken on with amazing courage.

Mark Pascoe, CEO, International WaterCentre.

A very true and honest account of one man's amazing and selfless journey to highlight the plight of our major inlands river systems. Queensland Canoeing wholly congratulates this effort to sustain the waterways for future users.

Mark Priestley, Executive Officer, Queensland Canoeing Incorporated

Steve Posselt is one of those amazing people who will risk all for an issue. He doesn't just talk about the problems facing the planet, he goes and sells his business, gets out there amongst it, and draws attention to the crises that we face. In this easy-to-read camp fireside chat, Steve vividly highlights the dreadful damage that we have done to, and continue to inflict on our rivers. Highly recommended for anyone concerned about our environment.

David A Hood, FIEAust CPEng
Chairman of Australia's College of Environmental Engineers
Chairman, Australian Green Infrastructure Council

Prologue

THE WATER WAS COLD, BLOODY COLD. The Southern Ocean is like that. White water tossed me carelessly, a rag in a washing machine. My kayak rode the boiling surf, alone, hundreds of metres away. The beach was even further – at least seven hundred metres. It was going to be a long swim – clutching the useless, precious paddle. The Southern Ocean, home of the great white shark, had won the day.

Tumbling through the surf with one bad shoulder was a harsh reminder of an earlier accident. It happened a year ago, on an outback road: cartwheeling through the air at 100 kilometres per hour beside my twirling motor cycle, whacking the bike somewhere high above the ground, and then nosediving into oblivion on the lonely, sandy road. When Davo shoved his hand up my jacket half an hour later, the pain brought me back to consciousness with a shock as he recoiled from the mess of jelly that he touched. I had been lucky to escape from that one.

Another year before that, we had sold our business to allow me to paddle the length of the Murray Darling and explore Australia's water crisis first hand. More than thirty years as an engineer in the water industry had given me plenty of time to think about the way we manage water, and I had become increasingly worried.

I had invented a water control gate that avoided the problem of acid sulphate soils leaching into the river systems and causing massive fish kills. It had taken many years and tens of thousands of dollars to develop but when it was finished I thought, 'This is still fundamentally wrong.' After thirty years I had still not got it. To try to control nature in this way eventually bites you on the bum. I think it would be better to go back to the natural landscape and work our land use around that. Sometimes it is better just to admit that the way you're going is wrong.

Prologue

I grew up in Grafton in the Northern rivers of New South Wales in the 50s and 60s. The grass grew because it rained from time to time, the ground was fertile and there was plenty of sun. The cows ate the grass. The farmers milked the cows. The local factory put it into bottles that were re-used over and over again, and we drank the milk. That system of farming was probably sustainable but it wasn't sustained. We thought we were smart and deregulated the milk industry. Now, the cow bails are all falling down. There is no butter factory in Grafton. It has been replaced by houses. At the same time, the amount of dairying in the Murray Darling basin has doubled. We now take 700 litres of water from an ailing river to make just one litre of milk.

When I was really young I used to know where the Easter Bunny lived. Mum told me he lived in Truckie Nott's corn paddock. Now this is not life shattering information, but it confirms a very significant fact. Truckie Nott did not irrigate his corn. Like the grass for Grafton's dairy cows, it grew on fertile river flats, nourished by the floods and watered by natural rainfall. Now there isn't a corn paddock anywhere in the area. Financial manipulation decrees that it is cheaper to irrigate the arid inland to grow food and build houses on fertile farmland.

Now, the inland rivers are failing and governments plan to move Australia's agriculture north. As a professional engineer with a lifetime of experience in building what we have now, I'm driven to help avoid repeating the mistakes of the past. To see the river with my own eyes and talk to the locals first hand seemed like a good start.

A friend who had narrowly escaped death from cancer told me this desire to save the world would soon pass. 'Near death experiences fade away,' was his opinion. On the contrary, I recall my decision every day and reaffirm it.

Part One

The plan was simple. Take a wheeled kayak from Brisbane to Adelaide, travelling the full length of the Darling River and lower Murray along the way.

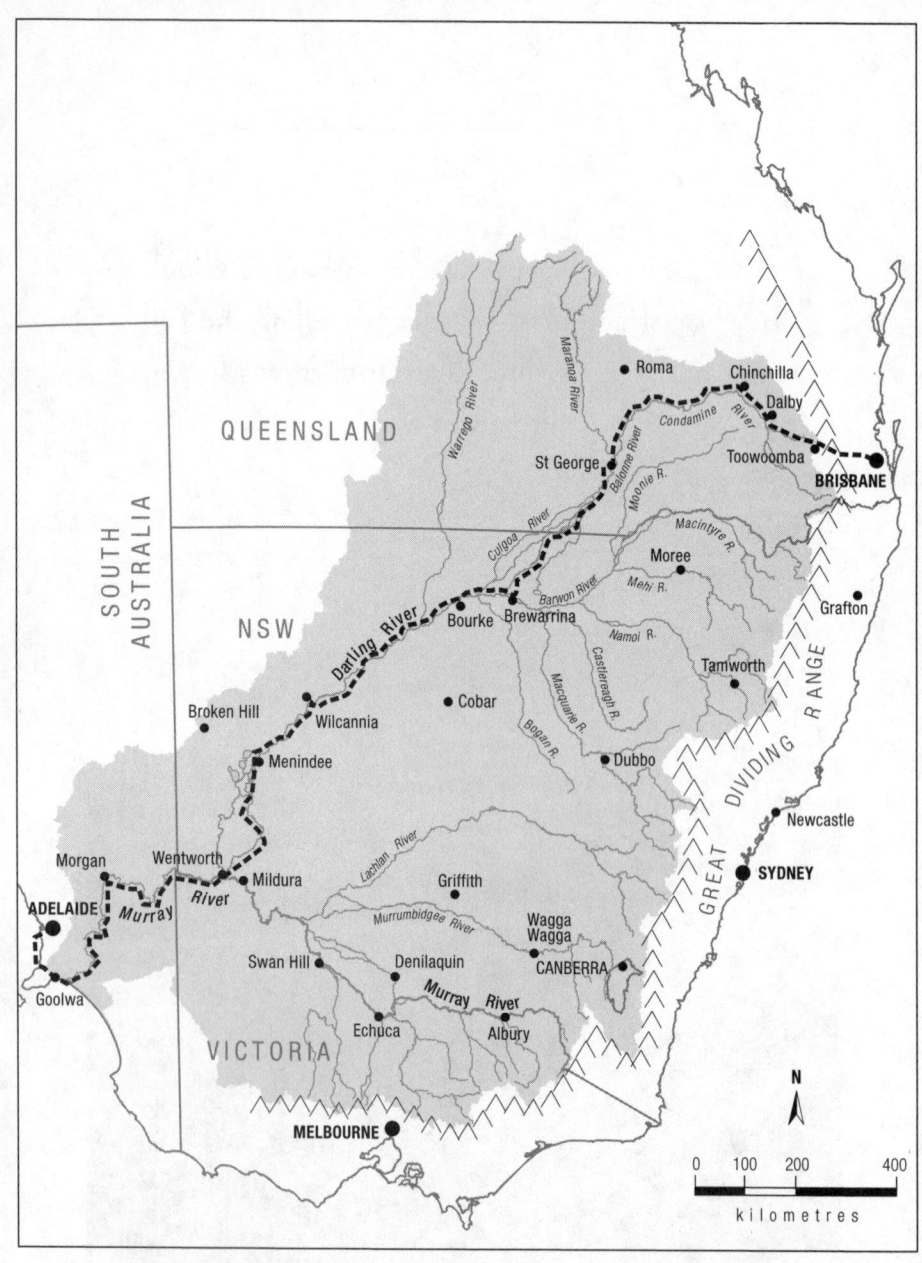

Steve's journey through the Murray Darling basin — 3,250km
2,170km paddling, 1,080km walking

CHAPTER ONE

An Idea Takes Shape

I knew my kayak. It was my oasis of peace as I paddled seven kilometres each way on the Brisbane River, commuting to work in a peaceful world of pelicans, ducks and fish, while thousands of poor souls battled the traffic roaring by on the banks as they headed to work.

I used the solitude to practise the art of grunting to pelicans. Pelican grunting is difficult to master, unless you're a pelican of course, and it only comes out right occasionally. I reckoned I had mastered the dominant male sound because they would fly away when I grunted. A kind friend told me that the ones that flew away were actually the females and this was the sort of effect I had on all women. Isn't it great to have mates?

Just on dawn one morning a duck flapped on the surface of the water as if it had a broken wing. Ducks do this if they have ducklings, to lead you away from the flock. 'I can catch this thing,' I thought. Full speed ahead, I got the kayak within three metres when, suddenly, there was no duck. There was a big swirl in the water as a bull shark took off with his catch. I hoped the duck had taken the edge off his appetite.

One morning I paddled an ABC radio announcer to work to promote Ozwater, the largest water conference and trade show in the Southern Hemisphere, which I convened in 2005 when it was held in Brisbane. He was worried by the shark story and amused by the pelican grunting. Setting out in the dark though, he was completely dumbfounded when a mullet jumped into the boat. The chances of that happening in his first five minutes on the river must be one in a squillion.

Paddling home from work one day, the idea of a great adventure occurred to me. I could take a wheeled kayak on a journey up the river from Brisbane, over the Great Dividing Range, down the Darling and Murray river systems and then around the sea to Adelaide. It was just a spur of the moment decision, but it seemed logical, and I tend to bring my ideas to fruition. Nobody had ever taken a kayak from Brisbane to Adelaide before. The fold up wheel arrangement we had developed was original and possibly unique. Hopefully I would come to understand the water, the land use and the people. The planned trip spanned three states and an enormous range of country, from the Pacific Ocean to the Southern Ocean through the river systems of the outback. The result would be a view of the water and land use from the perspective of a 54 year old civil engineer who had spent all of his working life in businesses focused on water.

The solitude of the river is great for thinking. At least two profound moments on my daily journeys extended the goals of the kayak trip. The first occurred after I read *The Weather Makers* by Tim Flannery. It hit me that it was madness to continue mining coal.

Both major political parties in Australia are trying to hoodwink us on 'clean coal'. There is no such thing; it is like a 'healthy cigarette'. The absolute holy grail of these guys – and they are a long, long way off it – is to mine the coal without releasing the huge amounts of greenhouse gas that we do now, then burn the coal in oxygen such that only carbon dioxide (CO_2) is left and, finally, to liquefy the CO_2 and hopefully secure it deep underground.

But even this simple, idealistic vision does not stack up. For a start, even in a liquid state the volume of carbon dioxide is five times larger than the volume of coal dug up and burnt! Despite this basic flaw in the logic, Queensland Government geologists are spending millions of taxpayers' dollars looking for sites to sequester the CO_2 in the ground. That money should be spent on renewable energy such as geothermal and concentrated solar. For every $10 of research money spent on renewable energy in Australia, $100 has been spent on nuclear energy and $1,000 on coal technology. This is back-to-front. At the very least we need a level playing field so that renewable energy is not disadvantaged in favour of vested interests that would squander these millions of dollars on unproven dreams.

The second epiphany, if I can call it that, came on an absolutely beautiful morning. The gum trees had recently shed their bark so the tall trunks glistened brightly in the morning sun. There are very few houses in the area so it probably looks much as it has always looked. Well, apart from a dirty river. I had a vision of the sea coming in, flooding back up the river, the trees dying and eventually being totally covered. I don't know whether it was the power of suggestion from what I had been studying, or a profound glimpse of the future. Wherever it came from, the image was powerful, disturbing, and returns occasionally to shake me up.

Some people regard my views on urban water as radical and some even regard me as a ratbag. As president of the Australian Water Association (AWA) in Queensland, I was on ABC radio in 1994 saying that we would be drinking recycled water in ten years. Although a committee of key industry players had been formed to build a demonstration plant, many people still thought I was a bit of a nutter. In the end I was only out by four years. By 2008, Queensland had developed the capacity to pump potable standard recycled water back up to Wivenhoe Dam so it can run through the system again. The Beattie government had not woken up to the issues until too late and panic construction was begun in 2007. People are still emotional about drinking recycled water, even though most people in the world do it without actually realising it.

After decades observing the water industry, I doubt that we can do our infrastructure planning in a sensible, intelligent way. The Queensland government has constructed a huge network of pipes around southeast Queensland to transfer water. A kilolitre of water, the amount you currently pay around a dollar for, weighs one tonne. To pump it 200 kilometres on flat ground uses about the same amount of electricity as running your air conditioner for an hour. Water does not move around as easily as electricity. There is a cost involved, both in real money and in energy.

Part of the problem is that people have the view that the water out of the tap is clear, pristine water captured up in the hills. Sure the water is clear and clean but it has been manufactured to that standard by the water supply authority. It would be an eye opener for most people to kayak down the Brisbane River from Wivenhoe Dam to the Mt Crosby water treatment plant. Here they would see kilometre long sections of water hyacinth blocking all use of the river. The hyacinth is rotting and stinking underneath, and transpires a significant amount of water over and above evaporation. They would see cows drinking, pooping and peeing in the water. They would see skeletons on the side of the river where animals have died and their bodies decayed into the water. For most people to connect all of this to what comes out of a tap is, I fear, an impossible task without a proper education program.

The details of Brisbane's water supply are so convoluted that they sound like a plot for a thriller. To get the gist of the story you need to know that water is treated using a six star rating system. The star rating system was devised by Jenifer Simpson. After a rocky start nearly twenty years ago, Jenifer and I have become firm friends. She's passionate and knowledgeable about water and recycling and, like me, does not care two hoots about the history of water. What we are interested in is its quality. Drinking water has five stars, six-star water is even purer. Any sewage plant on the river has to treat its sewage to at least three-star quality. That gives you a basic handle on the ratings.

Now the drama begins with the new recycling plants that take the output (mostly three-star quality) from existing sewage treatment plants and treat it to a six-star standard. That six-star water is to be pumped many kilometres to the Wivenhoe Dam, where it mixes into water that contains algae, nutrients, dead animals, and the urine and faeces of cows and other animals. It then goes to Mt Crosby water treatment plant where it is manufactured to five-star quality, using the same processes that have been in place for many decades, before it is pumped around Brisbane. Is this not just a tad illogical?

It is my observation that if we have to recycle the water after we use it, we are more likely to keep it clean. It is criminally wasteful to take clean water from a distant water basin, foul it and then tip it into the sea. That's why I am opposed to new dams such as the two new ones planned for southeast Queensland – Traveston and Wyalarong. These are conventional dams that take the natural environment or good farming land and cover it with water. They are part of the same thinking that whacks a city on what was once prime agricultural land. No one asks the question, what do you do as the city grows? Do you just keep on building dams, covering valuable farm land and bringing food from further afield? Where does it stop? When does it stop?

There is only one logical place for a new dam for Brisbane and that is at the mouth of the Brisbane River. The term 'stormwater harvesting' is on the lips of many. Frequently people will say that they see water running down the street and away and we should catch it. Where we could put it has not actually occurred to them. A barrage across the mouth of the Brisbane River would harvest much of the runoff from Brisbane. It would focus the whole city on the quality of what it allowed into the river. That way, everyone understands that the lower the quality of the water in the river, the higher the cost of treatment.

The environmental cost would be changing a salt water estuary into a fresh-water storage and restricting local flows into Moreton Bay. Given the state of the Brisbane River, this cost is not as high as some people might imagine. Already, the Brisbane River does not function as a river for much of the time, simply because there is often no flow past the Mt Crosby weir. The barrage could be placed far enough upstream to allow shipping to negotiate the Port of Brisbane and the river allowed to rise and fall with the seasons, instead of the tides. This sounds crazy to many people but it is exactly what Singapore does and is very similar to what I was to discover at the mouth of the Murray River.

The time had come to realise my dream of the great trip down the inland rivers. The plan was simple. Take a wheeled kayak from Brisbane to Adelaide, travelling the full length of the Darling River and lower

CHAPTER ONE

Murray along the way. It started as a light hearted comment: 'I might do that,' became 'Yep, I think I should do it' and progressed naturally to 'I'm definitely going to do it.' The trip evolved as if it had its own life.

The decision to take the trip led to other, tougher ones. First up, we had to decide if we could keep Watergates, the company that my wife, Carol, and I had created. After six years averaging 50 percent annual growth, it was a company on a mission. If I left it for six months there was a real danger that there might be nothing to come back to. The company was not mature enough to run itself without the owner manager being at the helm. We had to sell Watergates.

Peter Brown, a mate in recruitment and business brokering, found a buyer who could grow the business and look after the staff. I had made it tough for him because I did not want to sell overseas or to competitors. He later became a staunch supporter of the trip and even came out to Condamine with a couple of meals to die for. We had to compromise on the price because buyers value businesses based on average profits over the past three years, not on future growth. I remained as general manager but it was perfectly clear that I no longer ran the company. In the end I was depressed watching things go wrong, knowing exactly what was needed to prevent it, but being powerless to do anything. I had never been depressed before and I did not like it.

It was against this background that one of my mates, Geoff, organised an adventure that changed my life and the evolution of the kayak trip. He wanted to circumnavigate the Simpson Desert on big, off-road, BMW motor bikes. The plan included rough tracks, bulldust, Alice Springs, Uluru and a decent swag of outback Australia. I was in like a shot.

This would be my first trip with just the guys and no family since Carol and I were married more than 30 years ago. Carol would be busy with her father staying at our place while her mother had major heart surgery. At various times in our marriage it had been lucky for me that Carol is a registered nurse. There were quite a few times that she did the initial patch-up when I had pushed the boundaries a bit hard and finished up bleeding or breaking something. Heidi, our eldest daughter, would be back in Australia in between working-holidays overseas after finishing university. Amanda was busy with her doctorate on blue green algae and our youngest, Jonathan, had just completed university and was fixated on surfing. He was doing any sort of manual work he could to get the money to surf and party.

As it was to be a three week safari with plenty of action, we bought a commercial quality video camera. I wanted to practise using the video to record the kayak adventure the following year. At the Winton Hotel in far western Queensland we set up the movie camera and interviewed each other. Ken, one of the guys on the trip and a mate of Geoff's, did the filming and took to it like a duck to water. On reflection, our witty and insightful comments that night did not stand up to the harsh light of television. A video camera at a drinking session can be quite an eye opener.

More usefully that night, we started talking about my upcoming trip down the Darling River. Ken let on that he was taking long service leave from teaching and spending three months in the USA and Mexico. He thought he might extend it and come on the trip as the support crew and do the filming. And so a new piece of the plan fell into place. Ken was still keen the next morning and I decided to look no further for a crew.

It had taken three days to get to Winton but the next day was the first day in the dirt. Trialling the newly fitted tyres in a vacant block was my first experience with big knobbly dirt tires on such a powerful bike. The best bit, the really exciting bit, was the front wheel lifting up. I swear, the front wheel hovered about 50 millimetres off the ground – on the dirt. The Winton publican was a highly committed fundraiser for the Royal Flying Doctor Service. Swear words copped a fine of up to $20. Unfortunately I was probably his best customer that night. With the accommodation, meals, beer and fines my wallet was $200 lighter. Over a hundred bucks of that was for the Flying Doctors. Who knew, maybe we would need the service one day.

Out on the dirt road there were a number of mishaps, some scary moments and some low speed spills into bulldust pits. We agreed to ride well within our capabilities so I was travelling at around 100 kilo-

Chapter One

metre an hour. At that speed on dirt, the handlebars move around a fair bit. To travel across windrows of gravel they move a lot. It took me most of the morning to learn not to stiffen up and fight the movement. If you have to go through a soft bit you're supposed to stand up, drop down two gears, power on and let the bike do what it does best. At cattle grids the bike would leap into the air. The sound and the feel is of enormous, controlled power. At 120 kilometres an hour the feeling is hard to describe. I regretted the decision not to fit a helmet cam to show other people just what this bike stuff is all about: the speed, the sound and the three dimensional movement. Heaven! I loved it.

After lunch on 1st July we hit a long section of very soft sand. The tracks showed the other guys had chosen the left wheel track but I reckoned the right one looked better. About a hundred metres into the section, I found myself bouncing out of the wheel track. The wobbles were getting big, really big. I kept the power on because to shut it down would have been an instant crash. My job was to have a bit of guts and ride it out.

But it didn't work out that way. I performed a few pirouettes in the air beside 300 kilograms of motor bike, had an unfortunate collision with the bike high above the ground, and then knew nothing until half an hour later when Davo caused me to scream with pain as he felt around the damage. In real life Davo is a paramedic. I recall wondering why he had hurt me so much. He reckons I was frothing blood from the mouth and he had to confirm for himself how bad the mess was in there.

Two days later, I had been released from intensive care into the high dependency unit in Alice Springs Hospital with a positive pressure mask, oxygen and a drain into my chest cavity which filled a rectangular plastic container on the floor with the blood that shouldn't be there. The damage was extensive. All the ribs on one side were broken, some in a few places which means what is called a flail chest. That caused a haemothorax which means collapsed lung due to bleeding around it. I had a badly smashed shoulder, a numb elbow area, inner ear problems due to the bash on the head, and a lot of bleed spots in the lung. To keep me amused they gave me a bit of apparatus with three balls in it that I was supposed to suck to the top, but to start with I could not get even one of the balls to rise significantly.

Carol, my ever tolerant wife, had just flown out from Brisbane. She watched my feeble breathing for the next week and when I was finally able to ditch the drain tube from my chest, she rented a car and drove for three days to get me home because I could not travel by plane. We stuck to the bitumen and came back via Mt Isa. Even so, an emu ran into the side of the car, proving how unpredictable emus really are. We had worried about them on our bikes, especially as a mate, Monty, had nearly died the year before when his motorbike hit a bunch of them.

Back in Brisbane I did a tour of doctors and specialists including a chest specialist and David, an orthopaedic surgeon who specialised in major joints like shoulders. The chest bloke was easy. He reckoned that although my ribs were broken into little bits and my sternum was pushed around a bit that it would eventually mend and I would hardly notice it. That is, unless I was concerned about my looks. David's reaction to my CT scans was different. He clipped the images onto the light screen so that I could see them as well. They didn't look too flash to me. It was when he said nothing and just pursed his lips like he was saying "fff…fff…" that I started to draw my own conclusions.

At the next appointment, David had done some homework. He explained how rare the injury was, that there were six grades of severity, and mine was the second worst. He also said that any operation that had been done on these injuries had been a waste of time. We talked about my forthcoming paddle to Adelaide and he thought it might be possible in two years or even 18 months if I was lucky. Perhaps I protested too much but I remember him saying "Steve, you have to remember that this is a horrific injury." I thought that things were looking pretty good. We had gone from something like 'faaark' to 'horrific'. That's a definite improvement.

On reflection it is not that difficult to figure out why the injury is so rare. If you smash your shoulder into the ground hard enough to do that damage it is highly likely your head will hit very hard too. It's lucky for me that I have a thick skull.

Chapter One

A few weeks later David recommended a physiotherapist. I liked Darrin from the start. His was a 'stand up straight, have a bit of pride, don't look like a sook, and here's your list of exercises' sort of attitude. The exercises were pretty gentle at first because what little movement I did have caused me pain. It was, to coin a phrase, just what the doctor ordered. Like a poseur I wanted to impress Darrin with my improvement each week. With his magic and my stubbornness it all went very well.

Three weeks after the accident I decided to chuck the morphine habit. Even after such a short time I suffered withdrawals. I spent about four hours one night, hallucinating, sweating and feeling sick. I sure didn't want to start that habit again. It took six weeks to be able lie down in bed. Try that sometime. After weeks of sitting up you just dream about lying down. The mind is remarkable in how it just adapts to situations. With a pretty well busted body my frame of reference changed. To walk all the way up a hill near home was a great sense of achievement. As I started to get better I forgot how pathetic I had been and set new goals. It is subtle. The goals and expectations change gradually and afterwards I can only remember the big milestones. The time I managed to lie down flat for an hour or so was something to be savoured.

Towards the end of my physio treatment I rode what had been my second motorbike to the appointments. I did have some insurance money to replace the dead bike but there are certain difficulties explaining to your wife that having two motor bikes is a necessity. It looked like I would be a one-bike-bloke for the foreseeable future.

Black Betty is a 1600cc Kawasaki and looks a bit like a Harley with a big screen and leather saddle bags. The handlebars are quite wide, which I normally find very comfortable. Unfortunately, the first time I took her out was about three months after the accident and my shoulder was still not too good. It was all fine until I reached the front gate and went to turn left into town. My shoulder would not let my right arm go forwards enough to make the turn. Luckily there were no cars and no-one to see me mount the opposite footpath and execute some embarrassing recovery work. Next time though, I was well up to it. Out the gate, turn right, down to the roundabout and then come back up the street heading to town. Right turns were a breeze, it just took some planning to work out a route.

I still ride Betty any chance I can. She uses five litres of petrol per hundred kilometres. Even now, the best I have seen advertised for a small diesel car is about 5.6 litres per 100 kilometres. Because burning diesel creates more CO_2 than petrol does, this is about the same as 6.5 litres of petrol per 100 kilometres. The bike is still significantly less bad for the environment than the best small diesel. To do the maths for your vehicles, multiply the diesel figure by 2.7 and divide by 2.3. These are the relative kilograms of CO_2 produced per litre of fuel burnt.

Jonathan and I took my wave ski into the surf at Christmas time. Because we had spent a lot of time surfing together it was comforting to have him with me. Even though the surf was small and glassy I was frightened. I felt so weak and worried what would happen if I fell off or twisted my arm. But I was elated. It meant that I was on the road to recovery and could proceed with the big trip. I confirmed my decision to go ahead the following June.

In January I paddled 55 kilometres down the Brisbane River. Six months since the accident and the same time before I hoped to go, I felt for the first time that I was really going to be able to make it happen.

John Crocker, the senior boilermaker at work, offered to manufacture the wheel arrangement for the kayak. We used the shape of the hull to slide the supports on and we used existing attachments so no new holes had to be drilled. As our first design had leaked, this was a major consideration. We already had the flip-up arrangement for the front wheel perfected but the front pivot needed to be larger and stronger. When this was done the kayak did not track as easily and tended to have a mind of its own. This was solved by connecting the harness arrangement directly to the swivelling wheel instead of to the kayak. All weight had to be taken at the waist because my shoulders could not take any load. I looked a bit like a horse pulling a sulky, but it worked. My confidence was high. The dream was back on track.

Now I had to make it come true. Five months after the accident, in November 2006, I made presentations about the proposed trip to the

CHAPTER ONE

AWA regional conferences in Queensland and New South Wales. I talked about the trip, what I hoped to achieve, showed off the wheeled kayak and tried to convince them to help fund it. Our daughter, Amanda, and I started to develop a press release about the trip but we found this very hard and made countless attempts without getting it right. Part of the problem was that everyone wanted something different. AWA wanted the objectives in dot points, probably because the CEO was an engineer. Econnect, the company doing *pro bono* media support pushed more for the emotion behind the personal journey. Amanda was clever. She drew out all sorts of things about me and water that I had not even thought about. It was still a hard job though. Every time an Econnect professional revamped the press release it looked like too much bullshit to me.

Finally AWA was happy and agreed to commit. The press release was broadened to include more of what Econnect wanted and it was time to get a committee together. Geoff, who had organised the outback bike adventure, was keen to get into gear. Geoff is a joint owner of a company called MC^2 which had built the website for our business. Geoff and his wife Gail went to school with me. She's an English teacher so we called on her talents from time to time. He was happy to do the web site for nothing but first we needed to find a name for the site and the trip.

Geoff put forward a number of suggestions, many of which had come from Gail. We had a meeting with all the people who were interested in helping organise the trip. The meeting adopted Geoff's suggestion of Kayak4earth along with the K4e logo and he was off. He organised the domain name and started to build the web site. It is intensive work and I will always be extremely grateful for what he did. Most people do not see the hours that go into this sort of thing. At one stage I was really worried about how much pressure he was under with Kayak4earth on top of all his other tasks, so I rang Gail. She said it was fine; he was stressed but that K4e was more an outlet than an extra pressure because he enjoyed doing it. That was a huge relief.

It was interesting to observe the process of some people enthusiastically offering their support and then realising they could not afford the time while others get slowly drawn in. Perhaps the best surprise was the role taken on by Jenny Cobbin – making contact with all the schools along the way. Jenny had not even been at the first meeting but her husband, Grant, had kindly offered her services.

As part of my training I had attended a 'Leading From The Heart' workshop by Ozgreen and a shorter one called 'Despair to Empowerment' led by Ruth Rosenhek from the Rainforest Organisation. Ruth also introduced me to the wombat we used on the trip, but more about him later. Both workshops were very moving. The level of commitment and the depth of feelings were impressive and quite outside the professional character of the industry I was used to. I found it well worth doing this 'touchy feely stuff'. I know that expression is used as a put down and I could use 'getting in touch with your emotions' or 'learning to deal with emotions' instead. But 'touchy feely' accurately conveys the difference between how I usually approach things and this emotional side of life, people and the environment. Unfortunately, I will never be one of the touchy feely people, but I admire them and they are a great comfort and support before the next battle.

It was also time to chase sponsors. K4e already had the pump company KSB on board. Dave Alexander, who runs KSB, and I think alike and we have had similar successes in the water industry due to similar approaches. Dave and I both see the same things as blindingly obvious. It never ceases to amaze me that other people look at the world completely differently. I feel this way about global warming and fossil fuel burning. What is blindingly obvious and irrefutable to some, is never understood or is completely denied by many others. Even though we don't agree on everything, I always find it refreshing to talk to Dave: we even shared stands at the annual AWA regional conference in Queensland. KSB had more backing and more money but Watergates had a secret weapon: an outstanding aluminium drinks trough loaded with grog. It attracted customers in droves.

I had no idea what the trip was going to cost. The guy who did the first budget estimated over $300,000. That was not on. When we built up the Watergates business, our plan was frequently revised to avoid trouble or to capitalise on a situation. This trip was growing like that. I knew I was going to walk and paddle from Brisbane to Adelaide; expected there might not be much water in the rivers, and understood that more sponsorship would help us achieve more, but the detail was elusive.

AWA agreed to $15,000 in money plus $35,000 in kind. The 'in kind' meant that we were allowed to use employees' time to that value, but we hardly used it. KSB offered $10,000 and United Utilities from South Australia offered $5,000. Peter Brown, who had brokered the sale of Watergates, was looking for a corporation that wanted to promote sustainability but his letters were rejected. The big one never eventuated.

Jenny Cobbin and Grant, who had dobbed her in to contact the schools, own Epco. Their business works a lot with Simmonds and Bristow, a chemical laboratory and consulting company. The owner, David Bristow, offered Jenny $10,000. This was very handy. The huge chain BCF (Boating Camping Fishing) donated $3,500 in goods to help set up the camping gear for the support vehicle. Wandering around their store makes me feel like a kid in a lolly shop.

Peter Brown gave us $1,000 from his business, plus Watergates staff and ordinary folks contributed a total of almost $2,000. A local bloke called Farnie has a company called Global Satellite that specialises in fitting out 4WD vehicles with communication equipment. Farnie lent us a UHF handheld unit to match the one in the vehicle, a satellite phone and the first $50 a month of free usage. He and his wife have a son the same age as Jonathan. Carol and I had delivered Jonathan to Farnie's many times over the years, but had no idea what Farnie did for a living.

Sponsors were very supportive and keen on the trip. This was important to me. It gave us a feeling of legitimacy and the stickers they gave us to put on the trailer and the 4WD really made us look the part. With those stickers and the gear we had accumulated, we really started to look ready for an expedition.

CHAPTER TWO

Be Prepared, Be Very Prepared

THE TEAM AND I THOUGHT WE WOULD do climate change presentations to the general public along the way. We planned to start with a forum in Brisbane, one in Ipswich and then one in Toowoomba, with the rest to follow during the trip. This is one thing that we got very wrong. The Brisbane forum did not happen due to lack of organization. The Ipswich one went very well and the Toowoomba one luckily coincided with World Environment Day but can only be described as half baked. After that it was simply too hard because it was impossible to predict where we would be more than a week ahead. On the other hand, presentations to schools turned out to be a highlight.

Perhaps, given the complexities of organising forums, we should have predicted a low chance of success, particularly in towns on the road. The Ipswich Forum took a lot of time to organise and we were still working from the comfort of home. Speakers whom I wanted were not available. Others were new to me and I was concerned about what they would do. Eventually we settled on a half day forum, finishing at lunch time. But, after lots of organizing and discussion with speakers, there was not enough time in the morning after all. The list of preparations required seemed to be getting bigger, not smaller, and I needed to focus on the trip itself. Luckily for me, a friend from AWA, Vikki Uhlmann, is a professional facilitator and offered to put in some time. She took all the speakers' names and details from me, rearranged the order to flow a bit better and set about organizing it all with Ipswich City Council.

With Vikki on the job of arranging the Climate Change Forum I could get back to the big 'to do' list. It was now only a bit over a month before we were to set out. Two things that bothered me were a solar panel to charge batteries when I was away from the support vehicle and a means of purifying water in case I needed to drink from the river. Both of these were done at the last minute and so would have been ineffective, but thankfully it turned out that they were not needed.

The trailer was specially made for the trip and involved a great deal of detailed planning. It was a standard heavy duty unit except that it had an extended draw bar. Luckily, I had lots of trailer experience from the Watergates business. Rather than buy a truck, we had built a large trailer and towed it with a 4WD. This allowed Watergates to deliver quite big loads to remote locations without the cost of an extra vehicle. For a couple of years I was doing about 50,000 kilometres per year just in deliveries to Queensland, New South Wales, Victoria and South Australia. All this experience had taught me a lot about balance and trailer design.

We built the whole caboodle on a 7x4 foot box trailer, taking careful note of the regulatory requirements published on the RTA website for weight, maximum dimensions and overhang. The idea was to keep the total trailer plus load under 750 kilograms so that we did not need brakes. The trailer would also carry spare kayaks which are six metres long so we designed a frame to sit on top of the 7x4 box. This allowed the kayaks to be slung from the frame, thus keeping them well secured even when heat made them very soft. It also allowed us to add a cover and an awning that could be rolled out for camping. On the extended draw bar we had room for a rack that held seven 20 litre containers, the spare wheel and the barbeque plate.

Balance is very important in trailer design. You need good weight on the tow ball for driving but it is critical to be able to quickly get that weight back to zero in the event that you get bogged and need to disconnect the trailer and manoeuvre it manually. The kayaks on their frame set the weight too far back but 70 litres of diesel in a 200 litre drum near the front of the trailer, together with 140 litres of water on the front, gave us pretty good balance and the ability to shift it quickly by removing water containers.

As with all jobs built to a budget there are many compromises, but we got the major components right and during the trip Jonathan was to be quite grateful for the thought that had gone into it.

From about Christmas 2006 to May 27, the day we left, the preparations took at least 50 hours per week and most weeks much more. This was on top of my working three or four days per week at Watergates. Luckily for me, I had given up fighting the managers who represented the new owners of Watergates. That gave me more time and emotional energy to prepare to cross Australia on a kayak with wheels. Corporate life is easy if you play the game, even if it is a silly one.

Ken, one of the outback biking crew, had already shown ability with the camera and committed to coming on the trip after a holiday with his wife Barb in the USA. He had confirmed his commitment to be the support crew before leaving and it was agreed that Barb was welcome to come on the trip. They expected Barb to join us for about three months but maybe not go the whole way. It had fallen into place naturally and I was grateful not to have the separate job of organising a support crew.

On reflection, this was a major mistake. Isn't the role of a support crew to support the person, ie support me to get from Brisbane to Adelaide? I spent many hours poring over the detail of the equipment but simply trusted that people would all have the same motives and be up to the task. There is no doubt this was quite naïve. Even with the advantage of hindsight though, it is unlikely that I could have done anything differently. Building a trailer is something you can control in detail, finding people to support you involves many more variables, most of them out of your control. Sometimes you have to just let matters take their course and learn from the experience.

Chapter Two

There was much that needed to be done but there just wasn't enough time for it all. I had looked at electronic maps but could not find anything that would work in the remote areas. My conclusion was that we should stick with paper. At that time a mate, Bryce Jones, had some time on his hands and volunteered to take over the role of acquiring and organising the maps. It was great to have Bryce involved. We had been mates for 49 years, ever since we were five years old and growing up in Grafton, a country town of 14,000 people. Like me, Bryce had become a civil engineer and we did some growing up together in the early 70s as two engineering students from the country exploring the big city of Sydney with all of its temptations and pitfalls.

Bryce reported back that 1:250,000 topographical maps were what we needed. He would cut them into A4 pieces, Carol would laminate them and put them into ring binder folders. There would be an original and one copy. We met at Sunmap, the government retail map outlet, picked out the maps that we wanted and Bryce took them home. These maps took us just into New South Wales. Stan Boath from United Utilities agreed to get the maps from the New South Wales border to the mouth of the Murray on the Southern Ocean and send them to Bryce. The maps used a range of different scales. Some were 1:250,000, some were 1:100,000 and some were 1:50,000. This had significant implications which will unfold in the story down the rivers. There is no doubt that the decision to use the paper maps was the correct one. The fact that Carol laminated them was even better.

When we were at Sunmap choosing our maps I started to get a funny feeling. It was a bit like butterflies in the stomach, a bit like feeling sick. It was probably the first realisation of the magnitude of the task ahead – daunting, maybe even frightening, but at the same time exciting. I was keen to get started and find out what was out there.

Ken and Barb returned from their trip to North America in April and it was time to think about the requirements of the support camp and really do some shopping. This was when BCF came to the party and we picked out $3,500 worth of tent and camping gear, including a fridge, emergency beacon and global positioning satellite. To hold this gear, we bought two very large toolboxes and bolted them to the trailer. They had lockable lids so we could leave the trailer unattended without worrying that our gear would go missing.

Ken picked out the Global Positioning System unit (GPS) and it turned out to be a brilliant choice even though it was not until we got to Bourke that it was used to full advantage. I had bought an old GPS for sailing that did not have the features of the one Ken had picked. It could, however, calculate the distance from where you were standing to a set of co-ordinates you typed in. I'd have liked to learn how to use the new system but with a matter of days to go, there simply was not time. Ken played with the new one on a trip to Sydney and became quite proficient with it.

On a trip like this over large distances across remote areas, communication would be critical and 240 volt power would always be useful so the auto electrician installed an inverter, a car kit for the new Next G phone, a UHF radio and a VHF radio. The Next G phone with an in car kit and high gain aerial and a USB modem for the internet completed our affordable telephone and computer communication.

Carol transferred $5,000 to Ken and Barb's personal account to finance things they needed for the trip that they could not get as part of the generous donations from BCF. I was not sure how we would deal with this tax wise but all the expenses were legitimate and we were running the trip like a regular business. We had renamed our family company Kayak4earth Pty Ltd in the hope that one day it would earn enough money to provide a living.

Carol was not looking forward to the trip because of the long time I'd be away. It wouldn't be correct to say that I was in panic mode, but it was something approaching that. There was so much to do and so little time. Ken and Barb seemed to be excited and the rest of the team were looking forward to the start of something that we had been planning and working on for months.

Because Ken had shown such aptitude for filming on the motor bike adventure, he and I agreed that a major role of his would be to film the trip, including interviews with people on the way. The idea was that we would see what sort of film we could put together at the end. I told him he could have half of any profit that was made from the film. He embraced the role enthusiastically, taking the video camera I had bought the previous year to Sydney where two of his children worked in the television industry. He actually spent time with a film crew and learned a lot, including the fact that we had a pretty good camera with features that I had no idea about.

Geoff and I had also talked about the idea of getting some sort of financial reward after the trip. I didn't want to go back to real work and dreamed of part time consultancy to cover living expenses and a more full time business following my passions. Geoff was also looking forward to eventually being able do the things he really wanted to do. He enjoyed the web work and his work was impressive. We hoped we could figure out together how to make a living out of adventures like this, while making a difference in the world.

We had a very good video camera but we needed to have some way to back up each day's filming. We figured that not everything would need backing up. The tapes could be stored and sent home at regular intervals, but Ken could regularly download the good bits that he wanted to keep. Always meticulous, Geoff gave me a list of options and what he thought we needed. He's an Apple man. I hadn't been happy with my Mac experiences in the past but settled on a Macbook Pro with a 100GB external hard drive for extra storage. This would be for Ken as I'd have the secondhand Benq laptop we had bought to connect with the Telstra Next G broadband USB modem. Geoff was crucial to the success of the venture. He set the Apple up with everything we needed and instructed Ken in its operation.

When Ken came back from Sydney he had even been given a fold up white reflector that you use to get the right light or shade on the subject you're filming. Actually the right word there is 'talent', not 'subject'. From his working with a film crew Ken told us that when you go out to film someone they are the 'talent'. This means that you never have to waste time explaining to your mates or co-workers what you're doing, you are simply meeting the talent. "Hey Ken" I said, "I like that. If I'm the talent, I must be talented, eh?" Now why would he not concur with this analysis?

Bryce had taken the matter of safety very seriously and had undertaken a risk analysis of the trip. As manager of very large construction jobs he has no doubt saved a number of lives. His attention to detail, and his method of thinking, should be an inspiration to everyone. I appreciated what he did for us. The analysis was very detailed, but I couldn't take it in fully. What we were attempting had a huge number of unknowns. In my head I reckoned the areas of concern to me had been covered in my own way. There is an interesting character difference here. Bryce measures and analyses meticulously, exploring each 'what if' and recording it in his spreadsheet. My approach is to say 'Well we've no idea what to expect, let's try to make sure we have a number of backups to deal with the unexpected.' As for safety, I am concerned about it and my opinions are strong but they often differ from the concerns of others.

Bryce and his wife Jenny presented me with a safety vest with the K4e logo printed on it. The whole team signed it and I agreed to wear it whenever I was on the road. This was an emotional moment that brought the whole team together and even if Bryce and I have different approaches to managing risk, it meant a great deal to me.

With only three days before departure, Bryce and I drove the four and a half hours south to pick up Ken and Barb, as Carol was working. We stayed the night with Ken and Barb and next day did a trial run with the kayak, talking to children in the local school where Ken taught, which seemed to go quite well. Then it was back home to Karalee via Grafton.

Barb took me aside for our first serious talk about how we were to get on with each other in intense circumstances for such a long time. Foremost she said that dummy spits were OK, but whatever happened it was to be forgotten quickly and we would move on. I responded that dummy spitting was not what I do but that I understood, and as long as we could quickly get over things there would be no problem. We also agreed that they would be able to take days off and go sightseeing and that they would have two periods of about five days each, away, to honour some commitments they had. The first one would be on 5th July when they would travel to Sydney. We agreed that Carol would come out and stay with me on the trip and they could take her car. It was all reasonable and easy to agree to, so I didn't give a lot of thought to relationships. The banter between Ken and me was a bit blokey, I suppose. "What man wouldn't find a fart joke funny?" quipped Ken. Barb seemed to hold a different opinion.

Coming back we stopped at Grafton. This is where Bryce, Geoff, Gail and I went to school. To say it was a perfect place to grow up in the 60s is an understatement. I remember sitting out on the river with my mate at the time and reflecting on how lucky we were. That day we were on his VJ sailing boat, but most days we just went swimming or jumping off the bridge or floating on whatever was around at the time. I doubt there have ever been too many 15 year boys who look around them and want to keep their life the way it is forever.

The *Daily Examiner* is the Grafton paper and they asked to do a story on the trip. At the time, the federal government wanted to dam the Clarence River to supply a rapidly drying South East Queensland. The *Examiner* had a 'no dam' campaign, including a proliferation of stickers around the town saying 'Not a drop. Keep the Clarence mighty'. I have fairly strong views on the issue. I don't like projects that involve transfer of water from one river basin to another, then again, I don't really like dams. Maybe this is surprising given that Dam Design was one of my favourite subjects during my engineering degree. Balancing this, another favourite subject was Environment. I enjoyed learning about Environmental Impact Statements and delving into issues such as sand mining, flooding Lake Pedder and the growing impact of canal developments on mangroves and fish habitat. We were very lucky to be able to study this stuff in 1973. The foresight shown by the NSW Institute of Technology course designers is very rare today.

We continued our trip north towards home. The Clarence, Richmond and Tweed Rivers all suffer from drainage and associated acid sulphate problems. They are not the only areas suffering, it is also widespread further to the south. Near Port Macquarie we had designed and supplied a gate, a few years back, to hold the water level up in a creek to rectify the problem. On the Clarence I had given the matter a great deal of thought and study. In Cairns we finally got the design that I know will work. What you do is to install a swing type gate which lets water flow out and stays open long enough to let a predetermined level of tide back in, thus keeping the ground water high enough to prevent acid leaching from the soil. The swing arrangement, a bit like a door, allows bottom, surface and intermediate species to pass through. This is essential to get all the inhabitants in the marine ecosystem back up the drain.

It was reflecting on this design that made me decide there was something wrong with our whole approach to draining swamps. I had worked hard to refine a technology and minimise its environmental impact before I realised that the entire approach underlying the technology is wrong. Sometimes it is better to just admit that the way you're going is wrong, have a big re-think and start again.

Back home on the Friday there were now two days until our planned departure on Sunday 27th May around midday from Brisbane. We had a visit to the Karalee State School just up the road from our house to talk about the kayak and the trip and made frantic efforts to complete the trailer and gear set up. By now I was well and truly excited. Arrangements for the departure had been made, media alerted, and Lord Mayor Campbell Newman would be there to farewell us.

Stan from United Utilities came over from Adelaide for Saturday night and the Sunday send off. The team assembled at our place to have a barbeque. We got out the safety vest that Bryce and Jenny had given me the week before and anyone who had missed out before signed it. I don't remember much about the rest of the night, except being very careful not to drink too much and being preoccupied with what was to start the next day. Sleep was somewhat difficult but it was irrelevant anyway. Adrenalin would look after the next day.

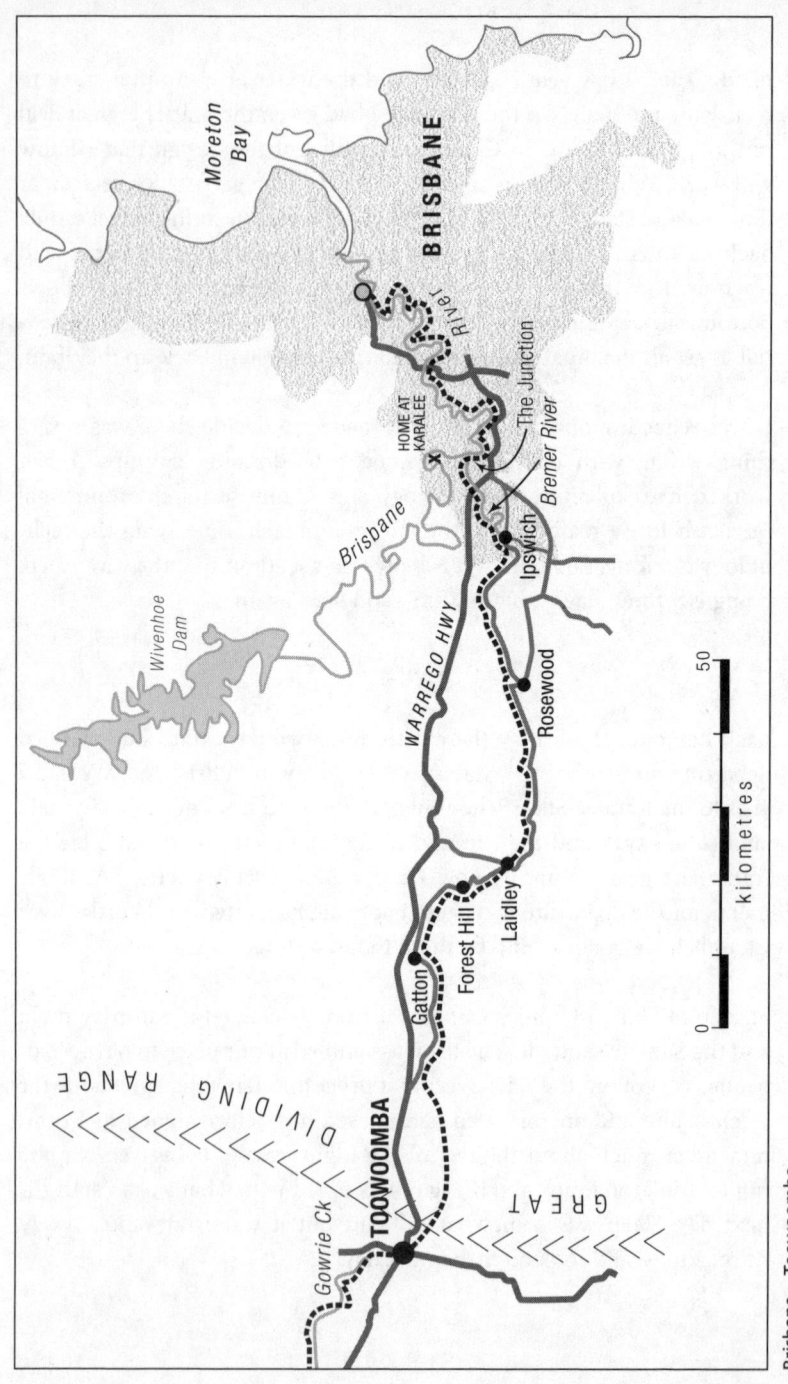

Brisbane – Toowoomba

CHAPTER THREE

Brisbane – Ipswich

27TH MAY 2007 – THE BIG DAY. We were all very excited. Brisbane's Lord Mayor, Campbell Newman, was to send me off at 1.00pm after a round of speeches at 12.45pm. The launch was to be at the West End boat ramp, a couple of kilometres from City Hall which was all boarded up during the construction of a new busway. I wanted to set out from Brisbane City Hall and end the journey at the Adelaide City Hall because it underlined the importance of the issue, the scale of the river systems and the links to the cities. That meant I had to walk from City Hall to the West End boat ramp before the official ceremony.

Carol, Barb, Ken and I left home around 8.00am with the loaded trailer and were the first to arrive at West End. Ken and I walked into City Hall and back - me towing the kayak and Ken filming. We had expected the sight of some bloke dragging a kayak up the mall on a Sunday morning to get a few stares but everyone took it in their stride. We were both in pretty high spirits. The whole trip was in front of us and we were excited about the start, the official launch and the media attention expected in a few hours.

Back at the boat ramp, people had started to arrive. Most were old friends but there were some very pleasurable surprises as well. It was difficult to see everyone, let alone have meaningful conversations with them. More than an hour before he was due, Lord Mayor Campbell Newman arrived. We talked about the trip. He had solar cells on his roof, as did Carol and I. He shared my understanding of climate change. He was keen for a paddle and looked pretty fit. In a flash we had the double kayak off the trailer and in the water, ready for him.

My new best mate, Cam, sat in the front with Bryce steering from the rear seat. Paddling alongside I joked that it might be fun to see a Lord Mayor tip over in the Brisbane River, but he had a compelling argument to the contrary. A wet mayor would be irresistible news footage. If we wanted to see our story on the television news, then Bryce had better make sure he stayed upright and dry otherwise we would be lucky to get a mention. Perhaps that is why he's a Lord Mayor and I'm a long distance paddler.

Our daughter, Amanda, was wrangling the media as well as introducing the speakers so had her hands very full. She did a great job and I'm very proud of her and extremely grateful that she was there to handle it all on the big day. She called everyone together at 12.45 and introduced the speakers.

First off was Chris Davis who talked about AWA's support of Kayak4earth. He quipped that there are a lot of 'Aorta' (They oughta) type people in Australia but he knew me to be a man of action. It's heartening to be recognised by your peers. Next up was the Lord Mayor who really nailed the issue of climate change and said many of the things that I was going to say. I was keen to get going so this was an unexpected bonus.

I thanked all the sponsors and supporters, terrified that I'd forget someone important but apparently made no memorable bloopers. Focussing on the definition of a sustainable activity as something that you can do forever, I said that climate change is only one of the things that's going to bite us very hard on the bum. I argued that all the problems that we face are due to our inability to live sustainably. We are taught that economic growth is essential and that we should consume resources to power the economy. How can an economy be independent of the environment? How can anything grow forever? But that is what we are led to believe. Endless economic growth is supposed to drive us.

With speeches done, the media became secondary. The trip had started and nothing was going to stop me. The long trek from one ocean to another, across the Great Dividing Range and down Australia's largest inland river system, had begun.

Amanda and her colleagues from Griffith University, where she was doing her doctorate on blue green algae, had arranged for me to use a probe to check the river quality between Brisbane and Ipswich. Surprisingly, even on these suburban waterways there was no continuous data of water quality from one end to the other. The probe was about 75 millimetres in diameter and went about 200 millimetres into the water. We had made a bracket to bolt onto the front wheel support frame. With the probe dragging in the water the kayak was bloody slow. Ninety percent of my total power gave only half the normal boat speed.

For the first hour and a half the tide was against me, going out to sea as I paddled up stream, making progress even slower. Bryce and Ken paddled the double kayak and stayed alongside for about 15 minutes of filming. An old canoe polo mate, Donald Leigh, stayed for a while cruising beside me in his polo bat – that's their name for the little kayak. Canoe polo is a bit like rugby on water. It is great fun, but I struggled with the rules and always seemed to be in trouble for ramming someone at the wrong time or being tipped over. On one memorable occasion I had been pushed over and rather than rolling back up, I climbed out under water. I surfaced into the middle of the players who were busy chastising the 'bully' who had tipped me into the drink. "You could drown the old guy," they insisted. She was half my size and about a third my age. I understood my place after that. You can always tell a polo player or a white water paddler by the way they use their paddle. Regrettably no one has ever picked me as either.

Another group that has played a significant role in my past is the Hash House Harriers. Alternately described as a group of runners with a drinking problem or a group of drinkers with a running problem, the Hash House started in 1948 in Malaysia. I was introduced to it in 1977 in Benghazi, Libya. To keep it all totally egalitarian every 'hash man', as we call ourselves, has a hash name. I'm lucky, mine is Knee Pads. Plenty of hash men with less flattering names pop up in this story, usually at just the right time. Monty, the bloke who had crashed his bike into the emus, had contacted a few hash people about the trip. One of the hash men, Sperm Whale, turned up in his tinnie with his wife Marion. Whale and Marion cruised alongside for two hours. Having the comfort of someone to talk to while the adrenalin seeped out of my system was fantastic.

Whale is infamous for his maritime adventures. Not content as a captain of industry, he sold his house to become captain of the boat that he lived on. Never one to do things by halves, he had also sat on the committee of a prestigious yacht club in Brisbane. Whilst in this illustrious position he achieved a unique honour by running smack bang into the most well known beacon off Brisbane's Manly Boat Harbour. His story is that it had something to do with the autopilot. The boat has gone now and he has a real house again, plus the tinnie and a campervan for extended outback trips. Little did I know that day on the Brisbane River, the hash lads would play a pivotal role in about six weeks' time.

When getting ready for a big trip you do the things that you have to do and hope that the small things will fall into place. There is simply not enough time to do everything. Learning how to operate the old GPS was one thing that had been overlooked. We had bought a new GPS and Ken had learned to use it on his trip to Sydney. He had velcroed it to the dash of the support vehicle to trace its route as he went. That meant I had no GPS for the river trip and was just guessing distances. At the time I thought it a bit back-to-front to track the support vehicle instead of the kayak, but there were a million other things to worry about.

With the adverse current and Amanda's probe slowing progress I looked forward to the tide change. A welcome call came from the crew suggesting we meet up river for lunch and a drink. It was handy to have the mobile phone in a waterproof bag on the deck of the kayak. Even in these first few hours, I was reminded of the different worlds on the banks and on the river. Given the extra effort and slow progress, I was keen to paddle as short a distance as possible. This means cutting every bend tightly unless there is a current further out that can be exploited – and that certainly doesn't happen when going against the tide The crew was not aware of this and arranged to meet me at a spot on the wrong side of the river. Despite that little annoyance, I had a welcome drink and a bite to eat and set off again about 3.30pm. Whale and Marion said goodbye and I wished them well.

Darkness fell and I began to see strange lights around the kayak. Eventually I realised that the probe lit up every time it took a set of readings. It's amazing for an old timer like me to see readings like these taken digitally, on the spot. Digital readings don't measure everything and laboratories are still essential, but a probe like this sure beats collecting regular water samples and taking them back to the laboratory. The parameters the probe tested were:

- **Salinity** - a measure of the salt concentration in the water. It is measured here in parts per thousand (ppt)
- **Conductivity** - a measure of the ability of the water to pass an electrical current. Here it is measured in milliSiemens per centimetre (mS/cm)
- **Total Dissolved Solids** - a measure of dissolved minerals in solution. It is closely related to conductivity and salinity. High levels mean lower water quality. Here it is measured in grams per litre (g/L)
- **pH** - a measure of the acidity or alkalinity of the water. Extremes of pH (less than 6.5 or greater than 9) can be toxic to aquatic organisms
- **Total Dissolved Oxygen** - a measure of the total amount of gaseous oxygen in the water. Levels below about 5mg/L (milligrams per litre) start to cause stresses to aquatic life
- **Temperature** - water temperature can be affected by many things (such as season and time of day) but generally it increases toward the lower end of the catchment.

The paddling was starting to get serious. I called in to home that they should pick me up for the night at the Goodna boat ramp. It wasn't until 8.15pm that I arrived. This was seven hours after the start, covering just 36 kilometres at best, compared with 50 kilometres in less than five hours over the same stretch of river a couple of years previously. The crew and their lights were a very welcome sight. That night was my first practice at sleeping like a log.

Sunrise, the national television breakfast show, wanted to talk to me the following morning and we had to buy more equipment and sort out the trailer. Ipswich was still about 25 kilometres upstream from where I had got out of the river last night. This included paddling past the very familiar Moggill ferry and then a left turn into the Bremer River. For obvious reasons, I was rather keen to do this on the incoming tide, which didn't start until about 4.00pm. We spent most of the day working on the trailer but still there seemed to be a lot left to do. We began to wonder if the job would ever be finished.

As usual the river was its own delightful world and the paddle very enjoyable. I was used to the slow progress caused by the probe. My hands were not used to the extra effort over such a long time but the gloves were preventing blisters. The sun was setting, the water was glassy, there was not a sign of civilization on the water and the only concern in my life right then was meeting Ken before dark at the Moggill ferry where he was waiting to film. Describing the idyllic setting to ABC Radio in Toowoomba held me up while mobile phone coverage was good, and the announcer seemed suitably envious. There were no employees, no customers, no one around and no cares. The way I felt at the time has a name. It was bliss.

The river layout didn't exactly conform to my mental picture. I'm always optimistic about how far it is to get somewhere. Darkness was closing and when I expected to be at the ferry there was another straight. 'Better extract the digit,' I thought, as I powered on until eventually coming round the bed to the ferry in quite bad light. Despite this, the quality of the film was pretty good. That is one benefit of spending a fortune on a video camera.

Pulling into the bank gave Ken and me a chance to plan the rest of the evening. Navigation was easy. I had a river to follow and the crew had a street directory. We decided that the destination for the night would be Cribb Park, just downstream from Ipswich's central business district. The plan was to keep our base at the house at Karalee until after we got to Toowoomba, except for a trial camping evening or two in the Lockyer Valley, somewhere near Gatton. This would allow all the outstanding items on our list to be completed plus anything else we found we needed after we tried the camping equipment. It would involve more driving with me being picked up every night and dropped back at the same place the next morning, but it seemed better than heading off not totally prepared. Carol was part of this decision as well. It would prolong the day that we would be apart at night.

The river near the ferry and The Junction, which is the confluence of the Brisbane and Bremer Rivers, was very familiar. This was the route to and from work. But tonight I turned left into the Bremer River instead of heading home up the Brisbane River.

Because of the extraction of water from the Mt Crosby weir, and the release from Wivenhoe Dam exactly matching the extraction, there had been no flow down here for a few years. Surprisingly, the ecosystem had come to life. Instead of behaving like a river, the area was now a long tidal inlet winding through the suburbs. Now people were catching a significant amount of mudcrabs and prawns. One bloke had even picked up two salmon. Two years ago, commuting to work on the river, I'd be lucky to see six people all year, now I'd see that many every week. It's a bit tragic when a river is better off without water flowing in it.

The Junction is a popular water skiing area. The water is muddy and many people subscribe to the theory that what you can't see, can't hurt you. Having seen some of the marine life taken out of that section of the river, though, you won't easily get me into the water there. It is still tidal, despite being 20 kilometres from the sea. One morning I chatted to a fisherman about his adventurous night. First he had been towed six kilometres by a bull shark before it broke away. Almost immediately after that, he snagged another one and managed to land it in the boat. As is always the case, the one he caught was nowhere near as big as the one that had got away, but he held it up anyway. He stood with both hands around the base of the tail. With his arms straight out in front of him, the head touched the bottom of the boat. That made this thing nearly two metres long.

Bull sharks are one of three species of shark known to regularly attack humans and they feed in estuaries in tropical and sub-tropical areas. Put that together with my duck story and you can see why I reckon the people in this part of the river are at risk.

A few kilometres upstream from The Junction, the Bremer River passes under the Warrego Highway and nearby is the Bundamba Sewage Treatment Plant. For years, Bundamba had provided the only significant flow of water into this part of the Bremer - although further downstream the meatworks treatment plant contributes a reasonable amount. Over the entire Brisbane and Bremer Rivers the probe showed no spike in any of the quality parameters we measured. That indicates there is no point source of major pollution anywhere. That doesn't mean the rivers are good, just that there is nowhere specific to point the finger at a polluter.

The Bremer was graded 'F' by Healthy Waterways. The South East Queensland Healthy Waterways Partnership was established in July 2001 and is a special collaboration between government, industry, researchers and the community. These partners work together to improve catchment management and waterway health. The organisation grades rivers on a scale from A for Excellent to F for Fail.

Interestingly, along the river near the Bundamba sewerage works there is a significant rise in dissolved oxygen indicating that its discharge increases the quality of the river. Bundamba is the first of Brisbane's sewerage plants to be upgraded for connection to the new recycled water grid. When I set out for Adelaide, those new works hadn't yet come online but it has to be good news for the Council and its operators.

Although familiar in the daylight the Bremer was confusing at night. A strong smell of coal gas which lasted for many kilometres indicated that the coal loader was near. I had helped with the construction of that loader after the previous one was destroyed in a flood maybe 20 years ago. The loader was trucked to the Port of Brisbane and placed on a huge barge together with a large crane. It then travelled for a day to get to its destination, almost back to where it had been built. The guys looking after the gear had a lazy day just sunning themselves on the barge. Everyone was happy, including the customer who had a new loader up and running in double quick time. Little did I realise what a significant role coal was to play in my thoughts 20 years later. How many people, even the locals, understand that there is an active mine in their midst and that the coal is barged down the river to the port?

The smells, the view, the guessing where I was had immersed me completely in my own world. The river does that to you. You no longer feel like you're part of the society raging above you on the banks.

Arriving at Cribb Park at Ipswich around 8.30pm I had a very sore arm so was pleased that this would be the end of pushing that blasted probe through the water. Ken picked me up and we went home for a beer and a sleep.

The next day was easy. The Brisbane and Bremer Rivers were now done. We had some time in the morning and used it to pick up the satellite phone from Farnie. Without him we wouldn't have had one so we were all most grateful.

It was about a one kilometre walk from the river at Cribb Park into the mall in the centre of Ipswich where we would spend a couple of hours meeting shoppers and talking to them about climate change issues. I was on the phone doing an ABC Radio interview for most of the walk in. We arrived right on time and set up the vehicle and trailer. It all looked impressive and people were talking to us in dribs and drabs. Some were extremely interesting, like the truckie who had travelled a lot in western Victoria and was most concerned about the drought and the issue of suicides which he had seen first hand. It was here that we learned about Bill. Bill was an older farmer up near Toowoomba and we arranged to go and see him and his creek on the way through.

By three o'clock we were back home and it was time to get my presentation ready for the forum the next day. Still at it the next morning, trying to get the PowerPoint slides finalised, I realised I had cut it too fine. As a result, I didn't speak as well as I would have liked.
The speakers whom we had arranged were:

- **Climate Change Overview**: Andrew Zuch from the Queensland Climate Change Centre of Excellence
- **Al Gore plus more**: David Hood
- **Renewable Energy**: Susan Jeanes, CEO Renewable Energy Generators of Australia
- **Wind Energy**: Lloyd Stumer, MD Wind Power Queensland
- **Sustainable Water for Ipswich**: Colin Hester, Ipswich City Council
- **Water Industry Strategies**: Chris Davis, CEO Australian Water Association
- **K4e Expedition**: Steve Posselt, Kayak4earth

Vikki Uhlmann was the facilitator and did an excellent job. It went to time, it was relevant and Vikki successfully brought it all together with some take home messages on what people should do.

Andrew Zuch had a couple of important slides that I used later which refer to the effect of average temperature on the top and bottom extremes. An average change of less than a degree seems small, but it has a dramatic effect on the number of cold days and the number of very hot days. This can have a huge effect on vegetation.

Lloyd Stumer had a slide showing that the number of Greenland earthquakes due to movement of the ice cap had increased by more than six times in the 13 years to 2003. We don't know the exact implications of this but we do know that if this ice gets to the sea either by melting or by sliding, sea levels will rise by about six metres. Knowing that the ice movement is doubling every few years is a compelling argument to take care.

Susan Jeanes listed the costs of various technologies and I was surprised to see just how high they are for nuclear and for coal with CO_2 capture and storage. Just how smart are we to spend the most research money on the most expensive solutions?

In a line up of great speakers the most inspiring was David Hood. He's passionate, articulate, comes at it from an engineering perspective and simply hammers home his sustainability message in a way that you cannot ignore. David points out the irrationality of what we do. In Queensland we have essentially discarded old building methods and have adopted a house design that looks like an oven with air conditioning. David is simply dumbfounded by the research direction taken by the Australian government and fears we will rue the day when we understand how much we have been left behind by those countries with more vision.

Ipswich City Council sponsored the whole thing, including lunch. Some delegates indicated that it was right up there with the best forums they had been to, but it was time to set off towards Toowoomba. The Mayor, Paul Pisasale, had been at a council meeting and so couldn't chair the forum. He arrived as I slipped into the harness. Amanda had been to school with Paul's daughter and played hockey with her. We had been to their home for a function before a school formal. This is the sort of thing that happens when you live in a place for a long time. The local daily paper, *The Queensland Times*, took a few photos, Paul gave me a certificate of appreciation and I was away at 2.00pm.

The next part of the journey would be all walking. I had paddled upstream from Brisbane to Ipswich and would now climb the Great Dividing Range to Toowoomba, 115 kilometres away. The pace that afternoon was fast. For two days we had gone nowhere due to the display in the mall and the time for the forum. I had read that the pace of an expedition should be built slowly but it was Wednesday afternoon and we needed to be in Toowoomba by Saturday night.

Chapter Three

It was four kilometres to Aquatec. This is a large water equipment company that I had been a director of until 1994, when I left to try something smaller. The staff lined up outside as we went past. There were lots of smiling faces, lots of old mates and plenty of good natured banter but there was only time for a quick stop. I knew this road as only runners can know roads. Training for a marathon at 39, I had run at lunch times, varying distances from six kilometres to 15 kilometres. All the turn around points for the various distances were very familiar.

The Bentleys live about 18 kilometres from Ipswich. We had worked together and known each other since 1985. Because the light was fading, I decided to stop there and leave the kayak with them. They have four delightful young boys who were interested in the kayak at first but quickly got back to the main task of establishing their own hierarchy. With the kayak safely tucked up on the veranda it was off back home again. I was buggered. The forum had probably used up nervous energy; likewise the socializing and meeting the mayor. On top of that, an 18 kilometres gallop and that was enough for the day. That left just 97 kilometres to Toowoomba but with a fairly steep ascent up the Toowoomba Range. In between there would be a nice little tester over the Minden Range to give the team an idea on how I was going.

CHAPTER FOUR

Up the Hill to Toowoomba

TOOWOOMBA IS A SIGNIFICANT MILESTONE for a number of reasons. Sitting on top of the Great Dividing Range, Toowoomba is the highest point on the trip. It marks the end of the journey out of the Brisbane water catchment and the beginning of the downhill journey to the mouth of the Murray River. Less than a year before my trip, the people of Toowoomba voted in a referendum not to drink recycled water. All the issues about managing and using water, transferring it from one catchment to another, treating and recycling it, come into sharp focus in this city.

From home it was about a half hour drive out to the Bentleys, where we had finished the previous night, and we didn't get there and into harness until after 7.00am. The Toowoomba Range was visible in the distance and I was keen to get a big day in. The city of Toowoomba was up there somewhere, nearly 100 kilometres away and we had to be there in just three days. The morning started cool but soon warmed up. I knew the area through Rosewood and up the highway but not as intimately as the section where I had done my marathon training. Walking through Rosewood most contentedly, watching people going to work and school, a dirty big hill coming up felt great – this was preferable to life in an office any day.

Late in the morning Carol came out to say hello. The cool drink, food and rest under the shade of a tree were interrupted by another radio interview. We were all very pleased with the coverage from the media. I managed to get my message about sustainability across most times and was learning more and more about how to conduct myself.

Just after Carol left to go to work, an old couple pulled up alongside. "How are you raising money?" they wanted to know. "I'm not raising money" I replied, "just awareness about the need to be sustainable". So we had a chat about life, waste, water and what we should do. They were from the Baden Powell Association and pressed me to take money. In the end I accepted $10 and said that it would go towards the projector and screen that we were going to donate to a school somewhere.

We were about five kilometres from Grandchester, a pretty little place blessed with not much more than a pub. "Hey Ken, I reckon we need to stop at Grandchester," I yelled.

Ken was quick on the uptake. "In that case I just may have a beer." Because there was still a long way to go that day it would be a lemon squash for me.

An hour later I pulled around the back of the pub amidst cheers and claps from a group on the veranda. Ken and Barb had driven ahead and let them know I was coming. The group had just buried someone who had been very close to them. It really is a good way to send someone off, to have a few drinks and talk about the good times. We don't really do death at all well in our society.

The conversation turned to the trip and what we hoped to achieve. These people understood the concept of turning lights off as you leave a room. The younger ones were concerned about climate change but an old bloke reckoned it was a 'load of baloney'. The drought was just a cycle and Premier Peter Beattie and Prime Minister John Howard were to blame. We would get to hear this often on the trip. Everyone has their own theory and the politicians get the blame for lots of problems. Today it was recycled water that was on people's minds. The Premier should have provided recycled water for the valley years ago, they thought.

The area is called the Lockyer Valley and it's the main food bowl for Brisbane. Crops need water and due to the drought many farmers hadn't planted that year. Everyone was concerned about the impact of this loss of income on the region. The discussion turned naturally to taking effluent from the Brisbane and Ipswich sewage treatment plants and using it for irrigation in the Lockyer Valley. It is quite likely that this had never eventuated because it couldn't be justified economically. It is very expensive to build water infrastructure and to pump water uphill. However, Brisbane's new recycled water system provides for some recycled water to be sent to the Lockyer Valley.

Despite the range of opinions the guys were friendly. A couple of them decided that I was a modern day Noah. I must know something that they didn't. Our drinks were free due to a generous publican and our companions gave us a lot of information about how people in the area were coping with the drought. We could have settled in for much longer but I had a significant hill – the Minden Range – to go over that day.

Along the relatively flat ground we decided to try the VHF radio. On the trip to pick up Ken and Barb, we had heard lots of chatter from boats on the car unit whenever we were near the coast. The aerial on the front of the ute was huge so we had high hopes that a hand-held unit in the kayak would be effective. These were dashed however. Maximum coverage handheld-to-vehicle was only one kilometre, which was pathetic.

Climbing the Minden Range was not too bad and we got to try Bryce's 'Caution Kayaker Ahead' sign. Barb would drive on ahead and place the sign on the road. The sign seemed to have the desired effect. Drivers would see the sign, go 'What the…?' and slow down for a good look at what could be a seriously deranged guy in a funny blue hat. Why would anyone be dragging a kayak on wheels over the top of a hill?

On the other side of the hill is Laidley. Coming into town there is a bridge with a sign that says "Warning: Diving or jumping off bridge is dangerous and prohibited. Rocks and snags under water surface". Just one problem, there was not a drop of water to be found.

In 2006 we installed twenty solar panels on the roof of our home. The photovoltaic cells can produce up to 3.15 kilowatts and we run it through a 5 kilowatts inverter to sell 240 volt power back to the grid. Just before the main street of Laidley I passed a solar installation company.

Further down the street, a young woman was sitting on the front veranda of a house with a baby and an older guy. I asked "Do you know a bloke called Warren who lives around here?" Warren had worked with us just after we started our business but had left because he was having trouble with the commuting. It transpired that I was actually speaking to Warren's wife, father and new baby. We arranged with them for Warren to catch up with us when we stopped for afternoon tea. Turns out he worked for the solar company. It's a small world sometimes.

There are two different ways to get to Forest Hill from Laidley, but I passed the first turn off without noticing, so there was no decision to make. A group of people on the veranda of a house all waved and wished me luck. A woman raced across the street, pressed $5 into my hand and insisted that I take it.

Ken arrived at a service station just outside Forest Hill as the sun was going down. The owners came out for a chat and brought a drink for each of us. They then filled the ten litre fuel drum for nothing. Ken and I loaded the kayak onto the vehicle and headed back to the Laidley caravan park. It had been a long day. We had met some very generous people, covered 37 kilometres and I had walked over a significant hill.

Rather than put my small tent up, I threw a tarp over the back of the kayak so that the canopy could be open and the mattress extension fitted to make it full length. This took less than a minute to erect and was very effective. Ken and Barb slept in the large tent that had been donated by BCF and they had rolled out the awning from the roof of the trailer. It was an effective and efficient set-up overall and we were all very happy with our efforts. I was asleep by 8.00pm which was to become the normal bed time for most of the trip.

The next day Ken dropped me back at the Forest Hill servo just as the sun came up and I set off at a fast clip because we were to stop at Gatton High School and talk to the kids at 9.00am. It was a crisp morning, fine and clear, with the Toowoomba Range looming ahead. This view was a constant reminder that the next day, Saturday, would be a tough one. Again it was fun watching everyone else on their way to work. It was really feeling like a proper adventure and thoroughly enjoyable.

Just before Gatton, I began to get a handle on the potential problem of road trains and big trucks. Always walking towards the oncoming traffic, my plan was that I could jump out of the way in an emergency. With six metres of kayak weighing 70kg tied onto me, realistically the chances were slim. A concrete truck was bearing down on me at about 100 kilometres per hour so the theory was about to be tested. When he went past I was nearly blown off my feet. Whoosh. 'Shit, he's blown my hat off.'

Turning around I couldn't see the hat. After I climbed out of the harness and walked back along the road, it was still nowhere to be seen. 'How can that be?' I thought. It was bright blue and pretty easy to spot. Still, it wasn't on the verge. It wasn't even in the paddocks 100 metres away. That hat was my favourite and was really a signature for people to recognise. It had sides to cover my ears and it was very lightweight. 'Bugger,' I thought, hopping back into my harness and setting off.

There were a lot of flies. They were buzzing around me and would land on my face and crawl into my eyes, nose and mouth. 'Bastards. Why don't they go and bother someone else.' They were even on my shirt front so God knows how many were on my back. 'Piss off,' I thought, waving my hands around. I brushed something on my neck. What's that? Well, I'll be buggered. It was my hat. It had blown off my head but the velcro on the flaps had kept it around my neck. It had been there all the time I had wandered up and down looking for it. I resolved to keep this little episode to myself. There isn't much point telling everyone that you're a dickhead when they can find that out in their own good time.

The journey to Gatton goes past a Queensland University campus. Heidi, our eldest daughter who was working in London, had done the first year of her degree there before the course was moved to the Ipswich campus. I had measured up for some gates in the drains from the pig pens a couple of times so the area was all quite familiar. A bloke on a bike stopped for a chat. He cycled from Gatton out to the uni every day, was very knowledgeable about water, climate change and supportive of the trip, so I enjoyed chatting. Time was marching on though and there were more than five kilometres to go in just under an hour to reach the school by 9.00am.

No sooner had he gone than a woman stopped her car. She had driven past, turned around and come back. "Ah Posselt, you haven't changed have you?" she called out. "You'll do anything to get on TV". It was great to see Leesa again. We had worked together at Aquatec about 13 years before and I had written her a reference for the job at the university. Despite the fact that I was now running late, we had a quick chat about life and where we were up to and then it was off again as fast as I could go.

Just where the road joins the main Warrego Highway to Toowoomba, Jenny Cobbin was waiting. Jenny was doing a great job contacting the schools and had driven out to see me try my hand at the first one. Throughout the trip Jenny set up almost all the school talks. Her efforts were invaluable. The kids were to become my greatest pleasure during the trip. They understood climate change and sustainability so I felt I could make a difference talking to them and connecting them to a wider movement. The teachers were already across the issues and K4e was just another tool that they could use to drive the message home.

Ten minutes before we were due at the school I was still on the old highway. Ken arrived and said there was nearly a kilometre to go so we quickly loaded the kayak onto the roof and drove towards the school. About two blocks away, we took it off and walked to where we met the principal and a journalist from the local paper, *The Gatton Star*.

The presentation was in a large covered area to kids who had been selected as the most interested. They certainly were keen. Their questions were probing and I enjoyed the exchange. Because we were in a high school, it was a bit more detailed about climate change and why 'clean coal' is like a 'healthy cigarette'. It started with the laminated map of Australia that Jenny Cobbin had given us. After five minutes I then said the lesson was over, it was time for the fun bit which was to talk about the adventure, the kayak and the paddle. I was not sure how much Ken and Barb wanted to be involved so I introduced them at this stage and then towards the end asked them if I had forgotten to mention anything. This left it open for them to introduce whatever they liked.

One of the useless bits of information I gave the kids was the effect that me towing a wheeled kayak had on animals. Dogs would look at me and bark. Cows would stare alternately at me and the kayak. Sheep had a bit of a look but lost interest rapidly. Horses looked at the kayak, not me. It fascinated them. They would stare at it and then race around rapidly, usually bucking and snorting. Then they would come back and stare again before repeating the whole performance.

Down the hill, at the old highway near where Ken had picked me up, there was a primary school. We decided to ask if they wanted to see the kayak. Ken, Barb, Jenny and the reporter from *The Gatton Star* drove to the school. There was an under eights carnival underway on the oval and they were keen. I walked down, going through the gate and onto the oval. The announcer introduced me, I said a quick hello and talked briefly about where we were headed. How many of those kids in years to come will remember the strange man with the funny blue hat and a kayak?

After that, the day was a bit of a drag. It may have been third-day-walking blues. It may have been the Range looming up ahead. It was certainly hot. My right Achilles tendon was bothering me. Also I was annoyed with Peter Beattie, the Queensland Premier. He had just issued a press statement about clean coal and how he was going to put carbon dioxide under the ground and ensure that it would be safe for hundreds of years to come. 'Great,' I thought. 'We have a Premier who simply has no idea.' Nuclear waste must be stored for something like 250,000 years while it decays to a safer state. Carbon dioxide is potentially dangerous forever. What use is a few hundred years? All this assumes we can find holes five times bigger than the coal we remove from the ground. Surely it is not acceptable to knowingly create such an enormous problem for future generations.

That train of thought had me going again. The papers were full of news about the Stern Report, an economic analysis of climate change produced for the British government by economist Sir Nicholas Stern. The report estimated the cost of the damage caused by each tonne of carbon dioxide emitted to be four times the price that Australia receives from the sale of its coal. If you earn a dollar in the coal industry – whether you sell the coal directly or build a mine or a train line, a road, or a port that services the coal industry – then someone, somewhere, sometime, is going to have to pay four dollars for you to earn your dollar. Where are the ethics in that? People justify it to themselves in all sorts of weird and wonderful ways. Even whole governments deny the facts. Coal is causing enormous damage to our global environment that costs much more to clean up than the revenue earned by mining it. There is simply no escaping that.

My frustration built up and I became angry with the world but it may have been just what I needed. It was Friday afternoon and I was 40 kilometres from Toowoomba which is at the top of a bloody great hill. On Sunday morning Toowoomba was hosting a World Environment Day event where I was to speak. Ken captured my mood perfectly. He was sitting on the tailgate, filming, while Barb drove and I walked along behind the vehicle. He wired me up with the remote mike so that I could be a fair distance back and he could use the zoom to get the frame set correctly. The video shows me tired, angry and getting stuck into politicians who will not address the science of climate change and cannot connect the dots. After my diatribe we were both laughing. "Yeah TEFE," said Ken. "You tell 'em." TEFE was a word that I had learned way back in 1976 that simply appealed to me. It stands for Tell 'Em Fuck 'Em.

After this, Ken and Barb drove on and found a farm where we could leave the kayak for the night. By the time I arrived it was dusk and I was exhausted. It felt like I was limping on both legs. Parking the kayak down between two sheds and walking back to the ute was a struggle. I folded myself into the back seat for the drive back to Laidley Caravan

Park and could almost taste the beer that was waiting there. By 8.00pm I had showered, wolfed down the dinner that Barb had cooked and was sound asleep in the back of the ute.

Saturday morning: the day of the big hill was here. Only 26 kilometres to the top, the morning's walk would be pretty flat and lunch would be at the bottom before the big climb. My legs had recovered overnight and I felt up to the task. We packed up the camp site, hooked up the trailer and set off for the kayak. We had spent two nights at the Laidley Caravan Park and the woman who ran it refused to take any payment.

The Warrego Highway, which is the main route up the mountain, looked far too dangerous. It was very busy and there was very little room between the white line on the edge of the bitumen and the Armco railing. We selected a smaller road to the south that was not very busy, but it was very windy and very steep for about two kilometres and quite steep for about another three kilometres.

Ken and Barb left me and set off for another look at the road ahead. When I approached the foothills they would operate the sign and try to warn motorists not to run over me. Every day I wore the yellow vest that Bryce had organised. The kayak is six metres long and bright yellow as well, so I had no concerns about being visible.

In these foothills I noticed something quite strange about myself. It took about two hours to catch up to Ken and Barb in the support vehicle after they had checked the road. I had been without contact for that time. When I approached the vehicle they watched me to make sure I was OK and then drove off to the next place to wait. Again, as I approached this would be repeated. There would always be some sort of comment on my lips; just smart-arsed banter, but nevertheless something to say. I never got the satisfaction of saying it, though. When I got the first glimpse of the vehicle it would become my next goal. The idea of reaching that goal was my mental reward. Before I could reach it though, the goal would move. I'd deflate a bit and give myself a stern talking to. 'Come on you big girl's blouse, get your act together,' I thought. Perhaps I should have suggested that they hang around for a bit so we could chat. At the time, I didn't want to let any human weakness hold me back but I'm not sure this was sensible.

In the end, it only made those occasions when I did catch up even sweeter. One of the things I had made for the trailer was a table that clipped onto the side tie rail. It had one fold up leg so it would always be stable no matter how the ground sloped. As I rounded one corner, Barb had the table out, a steaming hot billy and sandwiches for all of us. The coffee and sandwiches were great and the rest my legs got by sitting in the chair was even better.

The next time we caught up, Barb and Ken were standing by an ancient pump talking to the owner. The farm had been in his family for five generations and, during all of that time, things had never been so dry for so long. About seven years previously he had stopped growing tomatoes because there was simply not enough underground water anymore. It had been 15 years since he had seen a flood but before that they had been pretty regular – like nearly every year, he reckoned. Was this big dry due to a changing climate? "Well no. It's just a cycle and our records are only a hundred years old so who knows what happened before that," was his response.

Maybe the main issue was not climate change but over use. His farm was at the foothills of the Great Divide. Down in the Lockyer Valley, some of the aquifers that are the lifeblood of the food bowl have dropped alarmingly. Academics use the term 'water mining'. With any mine the product eventually runs out. The farmers are aware that they may be the last to farm the way they have been. What is to be done? One option is to split the land into hobby farms so the current owners can retire with dignity. This begs the question, how much land near major population centres can we afford to take out of service? With food and water shortages across the world, rising populations and the looming issue of peak oil, it seems obvious that we cannot lose any.

Bryce and Jenny turned up late morning. They had two friends with them and drove to Toowoomba for a coffee, promising to be back to help tackle the real pull up the Range.

On reaching the Range, Ken and Barb were in the ute behind me. We were about to find out if it is possible to drag a kayak up such a steep incline. The first really steep bit was a struggle. The front wheel lifted off the ground with each step. I literally struggled to put one foot in front of the other. Lifting my foot for the next step was no problem, but placing it further up the hill than the other one was extremely difficult. The 4WD following could stop anyone running into me, or me running backwards. Luckily 4WD vehicles have a low ratio gear option and Ken could get down to a very low speed and bring up the rear.

Just as the hill temporarily flattened to something more reasonable, Bryce, Jenny and their friends arrived again. Bryce took over driving the ute. Ken got back to filming and had his work cut out running ahead for the shots. My progress was so slow he had plenty of time. Halfway up we stopped for a rest and some food but the respite was brief. The local television cameraman came out and took some footage of me struggling but the main interview would be on the Monday.

Ken had me wired for sound. That means he could record remotely and hear through his ear phones. He was out of sight around the next bend. Certain that the really steep bits were behind us I rounded a bend and the road seemed to leap straight up into the sky. "Fark!" I said to myself. At least I thought it was to myself because no one was within earshot. When I got to Ken he was still laughing and said that I should turn the mike off. Seems like he had come around the corner, seen the steep bit and thought 'Steve's not going to like this'. The microphone proved him right.

When we had been filming on the river the previous Sunday, Ken had devised a series of signs to mean such things as 'wait I'm not ready' or 'film rolling, proceed' or 'come here I need to talk to you'. While I was going up the next steep bit Ken gave me a revision course. Then he said "What about this one", as he made a gesture with his middle finger. "Easy," I said. "That means my arsehole is dragging on the ground and I need to put it back." Ken and I found this a good deal funnier than Barb.

Bryce continued to drive the ute towing the trailer. Jenny drove their car and their two friends walked the whole way up, ahead of me. Everyone on the road wore a safety vest. A bit of a stickler for safety, Bryce had the system working perfectly with minimal interruption to other road users.

In all it was only a couple of hours of seriously hard slog. I was really pleased to get to the top but wanted to call it a day and leave the kayak with someone and go home. We were still four kilometres from Annan Park, where the World Environment Day event was to be held the following morning. We didn't have to be there until 10.00am so I had a couple of hours to cover the distance. Besides, Geoff and Gail would be at home and I was really keen to see them, especially as they hadn't been able to come to the departure in Brisbane. Geoff had settled into his role, uploading daily reports and photos to the web site he had created, which would be a huge task over a long time. Gail had been the one to suggest the name 'Kayak4earth' so these were two very special people.

At the top of the hill Barb asked some local residents if we could leave the kayak in their yard. They were pleased to help so I pulled it around beside the house. By the time the harness was off and the kayak empty, I was starting to feel like limping on two legs again. It was a relief to sit down in the ute for the trip home.

It was immensely satisfying to have the Toowoomba Range behind us. We had climbed 600 metres and now it would be basically downhill for a few thousand kilometres. It takes an hour to drive home from Toowoomba. My legs were stiffening up and my Achilles tendon was very tender indeed.

Ken was driving and by now he had covered quite a few kilometres with the loaded trailer so he was starting to get the feel of it. I have found that it takes about an hour to get used to the way a trailer performs. It gradually becomes instinctive and you find yourself easing up to the speed limit. We were doing about 100 kilometres per hour and about to go up the Minden Range. The road is dual carriageway. Ken overtook a car which was in the left lane. As he did so we started to go up the hill. With the trailer on we were slowing down so the car we were passing started to pull ahead again. "OK you C***," yelled

Ken, "I will play your game." He changed down a gear and managed to accelerate in front of the other car again.

I was sitting in the back and said nothing. I curse in my own way, but consider that language unacceptable and was stunned. Maybe this was an example of the dummy spits Barb had mentioned when she took me aside the evening Bryce and I picked her and Ken up. The fact that the language was extremely strong in front of his wife; the fact that the other car had simply maintained its speed; and the fact that it was obvious what would happen from the point when Ken started to overtake, all concerned me. Prior to that there had been no cracks in anyone's behaviour. 'A one off,' I hoped and put it aside.

When we got home I hopped out of the car only to discover a slight problem. I could not walk. The others went inside while I gingerly got my legs in motion and hobbled into the house. South East Queensland was on Level Five water restrictions and everyone was encouraged to have four minute showers. Normally a shower takes me less than this but I ditched any thought of a short shower and let the warm water flow down my legs. After a few minutes they started to feel normal. By the time I was dressed, I felt fine so it was out to the barbeque area for a few coldies and a chat with mates.

After dinner we just hung around the fire. It is a shame that almost every barbeque you see is gas fired. You will never catch me owning a gas barbeque. I take a lot of pride in my fires. The plate lifts up so that first we have a fire, then we fold the plate down for cooking, then we lift it up and sit around the fire. So, that's what we were doing. All of our children are familiar with cooking this way and Jonathan cooked for the night. Ken, Geoff and I were together and we started to talk about the ultimate goals of the trip. Geoff and I discussed making enough money to really raise awareness of the actions needed to reverse global warming. Ken was filming for similar reasons and I had agreed to split any film profits with him 50:50. That night, Ken, Geoff and I really felt like at team. Ken had become part of something we all wanted to achieve.

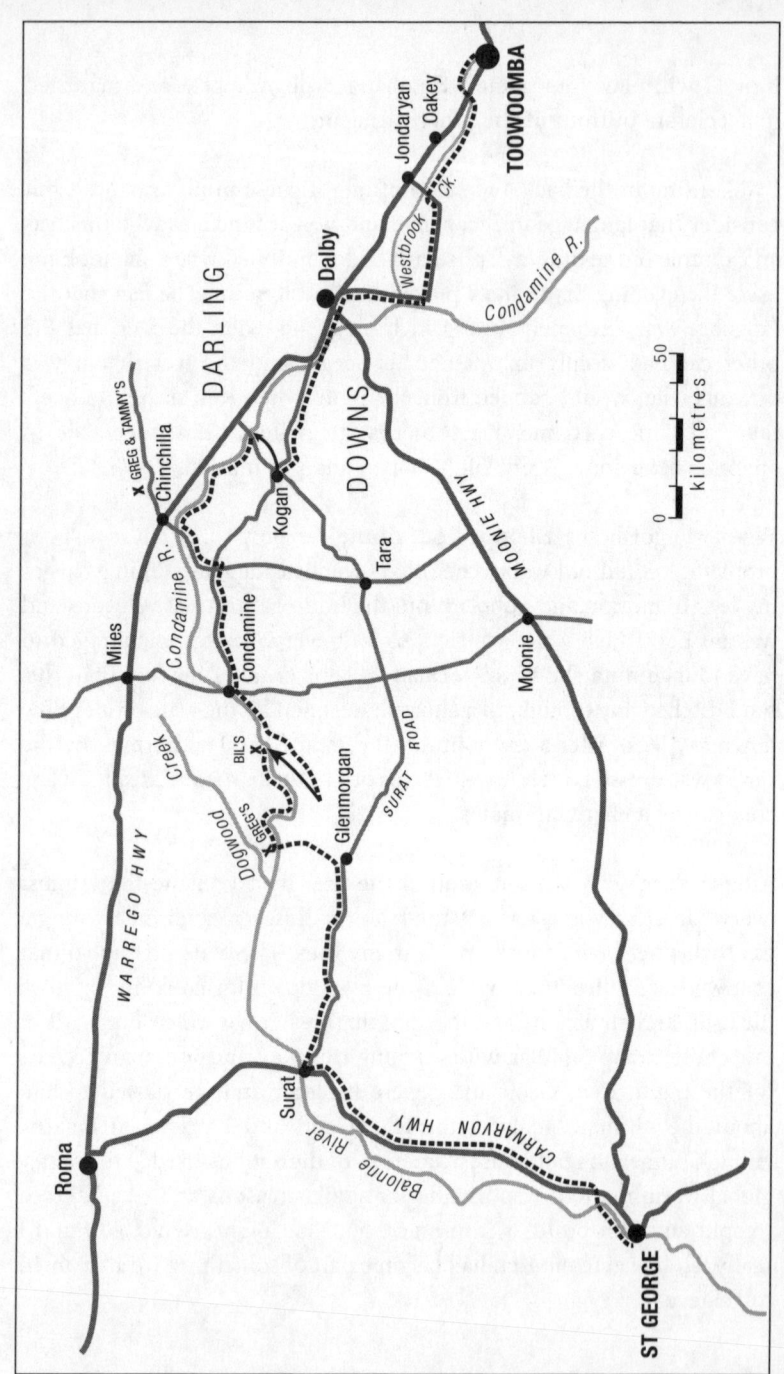

Toowoomba – St George

CHAPTER FIVE

Toowoomba

We were to meet Kev Flanagan at 10.00am at Annan Park. Kev, Toowoomba's City Engineer, had set up a tent in conjunction with AWA for the World Environment Day event. There would be a number of presentations on climate change, including mine. Running late, I called him at 9.20am to say we might not make it by 10.00 am. The team got to work quick smart setting up the ute, trailer, brochures etc. Carol drove me out to the kayak and Peter Brown came along to walk with me.

We covered the four kilometres in 35 minutes and arrived before 10.00am. Both Peter and I were pleased with that and agreed that we could not have done it any quicker. We arrived at the park amidst cheers and clapping. Our spirits soared. The soreness from the day before was gone.

Our little set up was kept busy talking to people most of the day. I finally got to meet Sarah Moles, the author of *The Dying Darling*. My friend Jenifer Simpson had put us in contact via email and now I got to put a face to the name. In 2006 Sarah and her husband, Michael, had made a number of car trips down the Darling River system. She interviewed people and Michael sketched them. Sarah's book explains in a very direct and human way some of the issues and problems that we face. The foreword to her book contains the following important paragraphs:

With climate researchers predicting a hotter, drier future for many inland areas, Australia faces some serious challenges in the 21st Century. Water is the most critical. Access to it is arguably the most pressing equity issue of our time.

Water reform is difficult. People get emotional about water – especially the lack of it. Managing water is highly technical. It involves a lot of maths, physics and computer models. Understanding it is hard work. Many people cannot be bothered trying.

Yet ordinary Australians, the majority of whom live in capital cities far from our major irrigation areas, need to be part of the dialogue about the role and value of water in inland Australia. The failure of Europeans to understand the 'boom and bust' cycle that characterises many Australian rivers, together with the over-allocation of water for human uses, has brought some of our rivers, their wetlands and floodplains, to the brink of ecological collapse. The Darling is among them.

I hoped that by the end of my journey I'd be able to add my expertise in managing water to Sarah's depth of understanding of the river systems.

Before my turn to speak, Professor Roger Stone from the University of Southern Queensland (USQ) presented his take on climate change. He had spoken at an AWA meeting at least ten years ago, saying that climatologists thought the drying of the south west of Western Australia was the first noticeable manifestation of global warming. Ten years later, Roger's presentation was on how the various parts of climate work together and will be affected by global warming. He emphasised that all the computer models predict La Niña events to dramatically reduce and El Niño to become close to the normal state. La Niña and El Niño are the two extreme states of the Pacific Ocean currents. The massive water movements of the world's greatest ocean control the weather from India to Chile. An El Niño event brings about reduced rainfall in Queensland and New South Wales and its opposite, La Niña, brings about increased rainfall in these areas. Roger noted that after 15 years of drought a La Niña was finally forming. Queensland had floods seven months after Roger's talk so his advice was correct. Other climatologists made similar predictions some months later.

Roger produced a powerful slide showing model predictions for average global temperature as provided by the National Centre for Atmospheric Research in the USA. His slide separated natural and man-made emissions, overlaid with predictions from the computer models and actual measurements. The period was for the past hundred years or so. It was amazing to see just how close the model was to the actual results. Not only can we predict future temperatures but we know what contribution each of the components makes. It is just one of the many tools available to demonstrate that global warming is real and what causes it. The hard bit is trying to predict just what this increased temperature means to life on earth. It is also important to remember that these predictions concern long term trends. The weather in any given year or on any given day cannot be predicted much in advance.

These models are incredibly important. For my engineering thesis in 1976 I redesigned the spillway for the Nepean Dam near Sydney. People today would laugh at the computer programming of the time. Everything was on punch cards and you had to wait all night to see the result of a small change to one line of code. But it was still fairly powerful stuff, based on what the engineers in the 1930s knew about statistics and hydrology, but with much greater power to crunch the numbers quickly. The engineers of the 1930s did not have many years of rainfall data, either. By the time I came along there were 40 years of additional data to analyse and the computers to do it. What this ultimately meant was that I had to double the spillway size to cater for the maximum probable flood. The extra data collected over those 40 years allowed my generation to make better predictions. And then along came climate change and all this was stuffed.

The climate models that factor in temperature change take in much broader data about the interplay of global systems. The example Roger displayed showed how accurately these models have predicted the trends. His main point is that the models, however limited, are better than simply assuming that things will continue as they have in the past. We now know that the reality of climate change is happening much faster than the models predict. Some effects of global warming, such as fires in the Arctic tundra, were not programmed into the model. This accuracy is crucial. The challenge is that they are less accurate converting that temperature into rainfall. There are still unknowns in the equation.

My presentation was again fairly average compared to what I expected of myself. I resolved to allocate more time to preparation in future. In the end I was glad that there were not many people in the tent and half of those were my friends. Friends can be very forgiving.

Kev Flanagan had some stories to tell, as did his Mayor, Di Thorley. They were prominent in the national media over their proposal to introduce potable reuse of Toowoomba's water. It was to be treated to six star quality and pumped to one of their three main dams. There was a public education and consultation process in place which had about two thirds of the population in favour of the procedure. Then at the beginning of 2006 along came Malcolm Turnbull, a relatively new politician and Prime Minister John Howard's parliamentary secretary on water, who insisted Council go to a referendum. Both Di and Kev believed at the time the referendum would sink the proposal. Some of us at AWA tried to help with information and education but Kev and Di were proved correct. The vote to secure Toowoomba's water supply by recycling was lost with only 40 percent in favour. The result reversed what Di and Kev had achieved by taking it slowly and leading the community one step at a time. The arguments put up by the opponents to water-recycling were sensationalist, fear-mongering and sometimes downright lies. The opponents won.

For political reasons, water authorities have begun to engage the population in the planning of many water solutions. This has only encouraged vested interests to sensationalise the debate. As a result, a lot of the public debate around these topics is an insult. It worries me that people with no understanding whatsoever can dominate the debate. It is critical that government processes are transparent but it is also important that experts are respected and left to get on with their job. We do not ask ratepayers whether they agree with the nine or so chemicals that are added to Brisbane's water every day. We do not ask them whether we should use chlorine, ultraviolet light or ozone as a sterilizing agent. We just give the population a manufactured water quality that meets the National Guidelines. The way that a complex issue is framed is critical to how the public responds. My view is that the history of the water is not the important issue, it is the final quality that matters. If the debate was framed in those terms a lot of the emotion would be dissipated.

Kev has an excellent presentation highlighting the tactics of his opponents. It is very funny – until you realise that the forces of ignorance won the battle.

After the day in the park the Kayak4earth team members each went our separate ways. Ken and Barb went off to stay with friends, Bryce and Peter got back to their lives and Carol and I went home in her Astra. When we arrived in Karalee, Carol and I visited Farnie at Global Satellite. He offered to lend us a UHF handheld transceiver that he suggested would be a lot better than the VHF unit. He was right.

On the way to Farnie's Carol asked "Do you remember Farnie's wife's name?" We had heard it once but it had slipped my mind. Then I recalled that it was Jenny. "How could we forget that?" I exclaimed. There was Jenny Jones, Bryce's wife; Jenny McDonnell, Peter's wife; Jenny Cobbin, Grant's wife; and Jenifer Simpson, indomitable water reformer: all part of our small team. Later we would find the same sort of thing where it seemed just about everyone we met was called Bill.

This was a short break. The following morning we met back in Toowoomba at the first of four schools for the day. Jenny Cobbin in liaison with Barb had organised the talks, so all I had to do was to go along and talk. First up was the Harristown Primary School and the kids were in grade five. That morning I learned a very valuable lesson. I had pitched the talk to what I thought was a grade five level of understanding. The kayak stuff is much the same whether the audience is five or 55 but there was a question on global warming that made my presentation look like kindergarten stuff. The question was very pertinent and showed a good grasp of the topic by the girl who asked it. I had no problem answering but I vowed never again to underestimate the children.

CHAPTER FIVE

We finished on time, barrelled across the road and talked to a grade 10 chemistry class. Because they were a senior chemistry class, I could go into more detail about carbon and oxygen atoms. Unfortunately the presentation was a bit ho hum because these older students were more interested in the kayak than anything else. Hopefully I could quickly develop the skill of picking what was appropriate for each group and there would be no more *faux pas*.

With a couple of hours up our sleeve Ken and I went off to Council to interview Mayor Di Thorley. As usual, she was very warm, unassuming and gave us a terrific interview. We had wired her up with the radio mike and Ken had chosen where she was to sit. This is what she said:

> I remember clearly stories from my grandfather who had come across overland from Adelaide, strange that now you are going there Steve. They worked for Jimmy Tyson sinking wells, never got paid if they couldn't find water, and then selected some land off Ballandean Station. Their whole life, of course, revolved around those creeks because if you didn't have a creek with a spring you could not get a well down to find water. I suppose they had a more symbiotic link to that water because without it you were gone, you would die. So they looked after the creeks. They made sure the trees were around the creeks, that the cattle did not go where the springs were because if you needed fresh water that was where you went. You protected that creek like it was your one life-line to everything else.
>
> But it seemed like in the 50s our whole world changed and people did not have to depend on that any more. There were tanks (for everyone) and then more and more people got onto reticulated water. And so we walked away from those creeks. We said we would get rid of the trees and we put little weirs in them and the creeks slowly died. The springs that were forever eternal, that you could just go and bury your face in, with the maidenhair fern growing along it, were no longer precious so you removed it all and stuck in a weir so you had water held back. We said we will change the way the rivers run, and we did. Mother Nature took a few years but She turned around and kicked us in the guts and said "I don't like what you have done to me". And maybe that's the biggest problem. It took so long that we as a nation and we as people who live on this planet just haven't got it. We can still

ignore it. You get a little bit of rain and somehow it will be OK, it is not such a big catastrophe, and we talk about the big droughts of the past. But we have changed things so drastically!

One example: They planted all the fruit trees at 24 feet spacing or whatever they were so that if it became dry they would survive, and then we got smart and we planted them close together so they were easy to spray and easy to drip irrigate, but when it comes times without any rain they can't survive. So our forebears knew how to do it to keep everything alive and they came from good old England where it was pretty much wet all the time. They came out here and did the same thing but we thought we were smarter than them. And what we did is we got so smart we started to destroy our nation, and we have done a good job of it. We have done a good job of thinking the water that goes past our property line is ours and we get as much as we can. We don't really think about the very heart and soul of our nation which is the rivers, and the connection between those rivers and what goes underground, what happens with that underground water. We have forgotten what it is like to see a river in its pristine state. We all talk about the money that we make to suck it out to grow cotton or oranges or whatever else but I wonder if we are really supposed to grow those products where the rain doesn't come out of the sky to allow you to grow them. Have we got so smart with our own technology that we have sown the seeds of our own destruction? I think we have.

When she finished I said "That was great, thanks Di." She unclipped the mike and handed it back and then proceeded to talk in more depth about her feelings. She's worried about the poor in a future world that is very different, one where water is much more expensive, where there are the haves and the have nots. She painted a clear picture of a possible future, a worrying future, if we do not watch out.

Essentially, what she said was, "Everyone is trying to predict the future based on the past and you can't do that." (Hang on, didn't Roger Stone say the same thing yesterday?)

"I'm just going to keep yelling. I have to look into seven grandkids' eyes and tell them I am trying to do something."

Di had spoken to some of the kids that we had visited. Her question to them was, "Are you going to be able to look into your grandchildren's eyes and say I remember this man coming to talk to me in 2007 and I decided that I'm not going to take this anymore. Are you going to be able to do that or are you just going to coast along and say, 'Yeah climate change is pretty bad eh.' This is what we have to achieve somehow, otherwise it will just get worse."

On the way out Ken chastised me for finishing the interview early. He pointed out that we had missed some golden moments. "You are not to finish the interview in future until we both agree," he said.

"You have me there mate," I concurred. There was no way that I wanted to miss such an important moment again. If you have it and you don't want it you can delete it, but if you don't record it you've lost it forever.

Now it was back to the park where Jenny Cobbin had bought some sandwiches for lunch. At 2.00pm a TV crew filmed us entering St Mary's School and then giving a presentation to Jeff Nolan's group. The previous May a group of us from AWA, Terry Loos, Jenifer Simpson and I, had run a one day climate change and water forum in Brisbane. Jeff had been pleasantly surprised by our depth of knowledge and had spoken eloquently at the summing up. Terry and Jenifer are both passionate about water, and would prove to be great sources of contacts on the trip.

The presentation was in the school hall and the group was quite large. It was interesting to watch Jeff interact with the kids. He was like one of them but not. They respected him and, more importantly, they had a thirst for knowledge.

From there it was over to the Rangeville School and a presentation under a tree on the oval. We must have made an impression because later we received letters from a lot of the kids about the trip and how they supported me. Some thoughts were quite practical such as the letter from one girl who said she was going to get her mum to buy me some toilet paper so I didn't run out on the trip. The kids seemed to like it when I showed the roll in a sealed bag and explained that I wouldn't fancy trying to use wet paper. The letters are tucked away now and are a prized possession.

Before she left, Jenny had a word to me about the importance of the schools and the support crew. Perhaps I hadn't included Ken and Barb enough in the presentations. It was incumbent on me to increase their role in these interactions where possible.

Despite having done four presentations, Ken, Barb and I were not yet finished for the day. We had an appointment out near Clifton with a bloke called Bill. We had been told about him at the Ipswich mall. Bill is about 70 give or take a bit. His family bought the property in 1867 and they have rainfall records since 1871, so we were keen to talk to him.

Bill's assessment of the rainfall over 140 years is that the average rainfall has not changed much in all that time. What has happened is that winter rain has fallen by about 10 percent and the summer rain now comes in quick storms rather than long soaking events. He sees that as a subtle drying of the country over the past hundred years or so.

What is much more significant is the impact of man. We were talking to Bill on a timber bridge. About a hundred years ago a man could ride his horse along the bank, under the bridge, and not have to duck his head. The local kids would dive off the bridge but only the very best could make it all the way to the bottom and bring up a handful of mud. When he was telling this story I checked out the bridge. It was well and truly past its use by date and under different circumstances I may have been worried it would collapse with me standing on it. But it was sitting on the ground. Bill was standing in the dry creek next to the bridge. His feet were less than a metre below mine. The creek has gone.

When Bill was young the creek here would fill about two days after a storm near Warwick. Now it takes about ten hours. Bill says the contour mounds are to blame. These are long mounds, about half a metre to a metre high, that wrap around the hills. He says they just funnel the water so everything just happens a whole lot faster. This is not what I thought contours did but when I talked to an ex Zimbabwean farmer, he confirmed that was how it worked in Zim.

Chapter Five

I'm not sure what to think about Bill and his creek. Maybe the creek was deep in 1900 because it had been scoured out since 1860. Maybe the level in 1860 was what it is now except that it was a series of swamps with lots of trees. I don't know but I'm willing to bet that it was not the shallow drain that it is now.

This was the sort of information we had come for. Sarah Moles had done a lot of research and had concluded that we had destroyed the river, so what else could we expect?

The day had gone exceptionally well and my body was well rested. We had covered a lot of ground, learned a lot, and maybe made a difference to some school kids. It was off home for our last night in the house. After this there was no going back.

When we got home Ken connected the Macbook Pro to the web and started uploading for the web site. I settled into getting my presentation up to scratch by adding slides from the key presenters. I was very much in climate change mode, excited by the excellent speakers in Ipswich and Toowoomba, and very pleased with the slides they had passed on. David Hood had motivated me with his passion at Ipswich. In all I was confident that we were up to date with the latest research and finally had a presentation that was good.

There was just one niggling problem. Ken had reacted extremely when Geoff asked him about the release forms for putting photographs of students on the Web. I believed Ken was blowing the issue out of all proportion. Even though the guys have known each other for decades and it didn't directly involve me, it was serious enough for Carol and me to discuss it. We had run a company of more than thirty people so were used to discussing strategy on how to get the best out of people. Carol's advice was to let this behaviour go but keep an eye on it. She said, "Steve, as the expedition leader, you have to do what it takes to jolly him along." I was reluctant to reward bad behaviour, and giving encouragement when it is not really due is not my way of working, but if that was what was required to make the trip a success then surely I could do that for six months.

Chapter Six

Onto the Darling Downs

THE LAST DAY AT HOME IN KARALEE meant all serious loose ends had to be addressed and finalised. Ken had purchased some soft foam to seal the tailgate of the ute. When the timbers were screwed on and the foam glued to them, covering every gap that we could see, the tailgate closed tightly with a firm back pressure. This was a relief because we knew that there would be a lot of dusty roads on the way and ute tailgates have many gaps.

It took six hours until we were all satisfied that enough had been done. At 11.00am we drove out the gate for the final time and embarked in high spirits for Annan Park at Toowoomba. After dropping me off at 12.15 Ken and Barb set about finding a caravan park for the night. Walking through Toowoomba I received lots of friendly toots from car drivers. This must have been due to the news coverage on the local television and the radio interviews. Ken and I tried the UHF radios and established how far apart we could be and still communicate. It seemed that two kilometres was the limit, at least in Toowoomba, so we stuck with mobile phone communication. Carol had arranged with Barb to pay her mobile phone bill so they were covered financially for this.

Barb called with the bad news that there were three large events in the area and all the caravan parks from Toowoomba to Oakey were full. Not a great start to our accommodation on the road! The good news was we had the option of staying with Barb's relatives, where she and Ken had

Chapter Six

stayed on the Saturday. This was a huge relief. Barb made the arrangements. We met at 2.30pm at the Wetalla sewage treatment plant that provides the first flow of water into the Darling River system.

Toowoomba is right at the top of the Murray Darling Basin. When it rains there, the water ends up 3,000 kilometres away in the Southern Ocean. Well, at least that's what it did before European settlement. Originally swamp land, the Toowoomba area probably provided long periods of flow out of these swamps. During dry periods now, the only flow in the creek is what comes out of the treatment works. At only 600 metres above sea level and 3,000 river kilometres from the sea, the river basin has a very gradual slope. The exception is the first few kilometres around the hills.

As a result of my 30 years working in water, a lot of sewage treatment plants are very familiar. Wetalla is no exception. I had been involved, on and off, in equipment supply for upgrades since 1985. We were welcomed to the plant and taken down to the outlet via the bottom gate. As expected, the only water in Gowrie Creek was flowing from the plant but I thought it might be enough to paddle on. This was somewhat optimistic. I only got 500 metres to the culvert at the next roadway before giving up and going back to the road.

Wetalla is a large and impressive plant. It has undergone many upgrades in both size and process over the years. After the big toxic algal bloom on the Darling River in the early 90s, water authorities made it a priority to improve nutrient removal processes. This had resulted in rapid improvement in biological techniques to reduce phosphorus levels. Wetalla was one of the first big plants to embrace the new technology. It is ironic that the people of Toowoomba voted not to take advantage of the opportunity to treat this water and return it to their dams. Plenty of other people downstream use it on the few thousand kilometres between Gowrie Creek and the ocean.

For some reason most people assume that the water is cleaner if it has been in the river for a while. In the industry we call it the 'magic mile' that people think will miraculously treat effluent water and turn it into drinking water. More surprisingly, I learned during the recycled water debate, a significant percentage of people want 'God' to have a hand in the process. That is why they think it is OK to put it into the river for someone else downstream to drink. An even greater percentage simply don't want to know.

About four kilometres down the road from the plant it started to rain. This was unexpected but very welcome. It also made me all the more pleased that Barb's relatives had generously put a proper roof over our heads. We bolted for the comfort of their house, via the local bottle shop of course, and arrived after dark with the rain pouring down.

It rained all night so it was a cold and wet morning when Ken and I set off at 7.00am. The radio said we had received 25 millimetres of rain. Out at Gowrie Creek the water was flowing strongly so paddling was possible. Ken was concerned about safety as the flow was strong and we had no idea what was down the creek. I was cold. The wind buffeted the bridge and the rain lashed down. A thrill of real adventure ran through me. Or was that just a chill?

Sliding into the river, the kayak almost tipped over. It is quite difficult to stay upright with the nose in the water and the stern still on the bank, a few feet above the water. At that angle, I couldn't use the paddle to maintain my balance. Caught in this compromised position between land and river, disaster could strike in a moment. Luckily I only dunked one arm and then the kayak was swept away. Ken yelled out from the bridge to be careful and watch out for barbed wire fences. My heart was in my mouth. I swept around the bend and was alone with no idea whatsoever of what I was about to encounter.

The bends were very tight and the kayak is six metres long. The banks were high, sometimes 10 metres of undercut black soil. At one stage the kayak was stuck on a corner with the nose wedged into a huge, soft, vertical bank. The current was rushing underneath and I was fighting to stay upright and twist the kayak around the corner. During this busy time I was not alone. A one metre long black snake joined me in the water. It seemed to be in about as much trouble as I was but luckily preferred the 10 metre high bank to the kayak. While attempting to edge my way out of my predicament I kept a close eye on it. Once free of the mud I whooshed off, leaving the snake to sort itself out as best it could.

Another tight bend and there was a fence – three rows of barbed wire above the water and a sheet of tin blocking half the width. There was no time to plan for much, two seconds later and I was caught, a human fly in a steel web. The nose had shot straight through the strands but the waterproof bag in front of the cockpit copped a large tear and then I got stuck – firmly clasped by one strand. To untangle myself, I had to grab the wire, between the barbs, with one cold, wet finger and heave the weight of the kayak backwards against the streaming water. It was not easy but the only alternative was to jump out and be in a much worse mess.

There were two options for getting past the wire after I untangled myself from it. I could put my head down and pass underneath with the risk of getting snagged by the neck, or lie back and risk having my nose shredded in the barbed wire. I opted to go face down. Just when I thought I was OK, a barb snagged on my upper back. I cannot get my right arm anywhere near that area since the accident and my left hand wouldn't reach either. Answer? Take the jacket off. It is not easy to get a jacket off while hanging onto the paddle as well as the barbed wire and then trying to contort your arms out of the sleeves while sitting in a kayak being pummelled by water and threatening to capsize. Just as a nasty pickle seemed inevitable, the fence let go. Phew!

Two bends later and again the kayak was too long to fit around the turn. Over the top of a little island it went – sideways. The island exit was a graceful roll upside down. With the wheels and my crook shoulder, rolling back up again was out of the question. I hadn't even been smart enough to put the spray deck on so the kayak was full of water. A spray deck, skirt, whatever you want to call it, is the plastic shield that goes around your waist and clips to the edge of the cockpit to keep the water out. I rolled out underwater and found the creek bottom about a metre down. The island came in handy as I was able to put the back of the kayak onto it, pick up the front and tip it over to empty it out.

Getting back in was easy so away I went again. All of the activity had kept me warm but the air temperature was about seven degrees and the water not much warmer. The next fence was fine as the farmer had hinged a piece of tin from one strand of wire. The kayak simply crashed through. The fences became less of a problem as the water got deeper and I could often shoot straight over the top row of wire. Many times the kayak slammed into the top of a star picket so I was very grateful to have a plastic boat. A fibreglass one is lighter and faster but would be torn to pieces by these steel posts.

After battling along for another hour I came to a causeway where Ken and Barb could come and pick me up. We had a presentation to do at the Oakey High School and I wanted to get out of my wet clothes and warm up. Life is not always simple though. Ken and Barb couldn't find me from my description of the causeway. Preparing to walk all the way out to the highway I saw something about two kilometres away. "Flash your lights Ken," I said on the phone. He did so. "Gotcha," I confirmed. "I'm about two k's straight ahead, on a bitumen road running down the hill from the Warrego Highway".

"Bewdy," said Ken. "One minor problem, Barb and I are stuck."

Ken had driven along the road in the right direction but it had turned into black soil. Black soil and water make a very sticky combination. Ken had stopped before he descended a hill. They opted to take the trailer off, turn it around and then turn the ute around. It was dirty, sticky work but they managed it and picked me up half an hour later. We arrived at the school five minutes before I was due to speak, and we were offered a hot coffee. I drank it while I took my wet clothes off and changed in a back room. Ken helped me lift the kayak onto the stage. The school assembly formalities were dispensed with quickly and it was my turn. I spoke the usual stuff about the kayak and the paddle, what the trip was about and what we thought about sustainability and global warming. My patter was getting better. I was also working Ken and Barb's role into the presentation a bit better by asking them to speak.

When the presentation was over I threw the wet gear onto the kayak seat and headed off down the road to the primary school. It was raining quite heavily and there was a lot of water about. Associated with the primary school was a pre-school building that was not being used.

There was no accommodation in town due to the three events in the area, so the headmaster offered us the use of the pre-school. There were also staff showers at the school that we were free to use. This was great, two days of rain, and two days with a roof provided by someone.

We were due to present at Jondaryan school about 20 kilometres away so we thanked the headmaster for his hospitality and drove out there. Jondaryan was a small school and the older kids seemed to take responsibility for the younger ones. Barb talked to them a lot more than in previous presentations and seemed to enjoy it. I was still cold but feeling more than ever that this was the start of an adventure. It was fun. There were still a few hours of daylight after we finished so we headed back out to Gowrie Creek.

Where the creek crosses the highway is a good example of different worlds intersecting. Gowrie Creek at this point is a scar cut down a few metres below a treeless plain. Above the bank is grass, the roar of the highway and countless examples of civilization. Down on the water, once away from the bridge, it is just the banks, the rushing grey water, and the kayak. The view is only as far as the next bend. From the land, one hundred metres from the bank, the creek cannot be seen so a kayaker just disappears into a world inhabited by no one else. That day there wasn't even any bird or animal life to keep me company.

The flow was good and the bends were a little longer so the kayak could get round in one go. Westbrook Creek came in from the left and it had more water than there was in Gowrie Creek. Life was getting better and a whole lot more fun. Where a farmer had built a concrete causeway, the water would build up behind it, cascade over a 600 millimetre vertical drop and then tumble down a series of rocks, dropping about another metre. This was exciting, not scary or stressful, just a lot of fun. The kayak would hit the concrete and the rocks but bounce off unscathed.

Willow trees were a recurring problem. They catch everything. All the hanging branches give the weeping willow its name but they also have masses of roots that hang in the water. These dense, tangled mats even grab sediment. A kayak with protruding wheels is easy prey so a lot of time was spent untangling my kayak from willow roots.

Before dark I reached the road where Ken and Barb were waiting. We loaded the kayak onto the trailer and set off for the pre-school and a well earned shower.

The following morning it was only 50 metres downstream before the first raft of logs barred the way. It was about five metres long and about 200 millimetres high. The only way to get over it was to paddle furiously, slide up as high as possible and then put my hands down and inch the kayak over the logs. It was hard work. This sort of thing continued. There were about a hundred such log jams. The nose of the kayak got stuck in the forks of trees; the wheels caught in branches; the bottom got stuck in the mud on the banks. It was a real fight to get anywhere and I only made what I guessed to be about three kilometres in four hours. In amongst all this Westbrook Creek had joined Oakey Creek but I had missed seeing the confluence. By midday Ken and Barb had caught me at a bridge and persuaded me that what I was doing was perhaps crazy.

They had been a lot more successful than me during the morning. The principal of the Oakey State School had introduced them to Uncle Bill, an Aboriginal elder whom the school had adopted. Never short of words, Uncle Bill is always happy to regale listeners with stories of his life. He's from Louth, which is downstream of Bourke on the Darling River. For the past four years he had been in Oakey painting murals and making school business his business.

He says that "White man has ruined everything, cleared land, pumped water, nothing running into the rivers." When he was a boy "there was always water and fish. Now no fish." And according to him "it is gonna get worse." As a young man he used to work some of the big cattle properties in western Queensland. Uncle Bill laments the passing of that era when it was easy for him to get work, when the stations were always busy. Many of the properties are gone: "Nothing there now," he says.

Just a few days before, Professor Roger Stone was predicting that a La Niña was establishing itself strongly. Now Uncle Bill was saying that the bull ants were flying, the long neck turtles were making their way up to higher ground to lay eggs, and the termites were building their mounds higher. He was certain that rains were coming. I couldn't help wondering though, whether someone should tell the long neck turtles that the huge pumps and the ring tanks (large farm dams) had changed a lot of things.

The town of Oakey lies on Oakey Creek. Apparently it is named after oak trees that lined the banks. Presumably these were she-oaks. We couldn't find any evidence of them now.

It was clear under the bridge where I met Ken and Barb, and easy to get out of the kayak onto a concrete block. Ken wanted to show me how to get in and out of a kayak but I prefer to do it my own way. Many people put the paddle behind them and hold onto it and the edge of the cockpit. That way if you lean towards the bank side the paddle holds you up. My problem was that I'd have no hope of supporting my weight on the paddle behind me. My right arm will not go there. In fact it took me five months from the accident until I could reach round enough to wipe my bum with my right hand. Try swapping hands for this important job one day and you will understand the problem. To get into the kayak, I hold the paddle in front of me with one hand and use the same principle. In truth, it is really not necessary. A sea kayak is a very stable craft compared to the racing kayaks that many people use. To get in and out, all I need to do is make sure my feet are in the middle of the boat and stand up.

I actually learnt to paddle properly, racing in a K1. I had already spent 20 years paddling a wave ski and reckoned I knew what I was doing. Wrong. On my first visit to a K1 club, the coach was suitably impressed when I paddled across the river, turned around slowly and returned triumphantly, only to blow it and fall out right in front of him.

For the next three months I could be paddling along comfortably and all of a sudden find myself in the water. The coach despaired over my lack of progress for a long time. "Slow down and get it right Steve," he would say. "You will get the speed after you get the technique."

One day it all clicked and the boat was going brilliantly. Being naturally exuberant, I felt that if I could paddle I might as well enter the state championships. You can enter in age brackets, so it is not as impressive as it sounds. I came last in every race and in the doubles we fell out before the start. If I ever do that again all I have to do is stay in the boat and it will be an improvement. At least I finally learned to paddle efficiently and powerfully.

When Ken and Barb picked me up I was exhausted. Ken's notes on the web site say that at that pace we could make Adelaide by Christmas, 2010. It was still raining so we headed towards Dalby where we had arranged accommodation. When we drove off there was a crash. I knew the sound and jumped out. Barb had placed the paddle on the roof racks and Ken and I had neglected to tie it on. It is a carbon fibre construction with titanium plates across the ends. One of the ends was damaged. It was still usable but would need some fibreglass repairs. Luckily, it was no big deal and there was a spare so repairs could be postponed.

It was Thursday and we booked into the Dalby Tourist Park where we were to do a presentation on the Friday evening. Dick, the owner, gave us a free night in a cabin so we paid for the Friday and Saturday nights. After we had booked in and unhooked the trailer, Ken drove me back out to the bridge at Jondaryan. While we were away the water level had dropped further so there was no point getting back into the creek. I started walking again.

Walking gave me time to reflect on the past two days and what I had seen. The creek had started to flow after rainfall of 25 millimetres. That increased to more than 50 millimetres. While it was still raining the water level was falling as I came downstream. There had been a couple of large ring tanks that stored water from the river but these couldn't possibly have accounted for all of the water. The only logical explanation is that this was an aquifer recharge area and much of the creek had simply gone underground.

I also reflected on the willow trees. There are different schools of thought on willow trees. Governments spend enormous sums to get rid of them after their predecessors had encouraged the planting of them. Some environmentalists plant them on their properties arguing that they do a great job of slowing down the flow until natives can eventually overtake them. Some think they are a scourge.

They certainly slow the flow down and grab debris and dirt in their roots and branches. Why were there not many on the outside of bends? It seemed to me that they were providing stabilization of the banks. They were on the inside of bends because they had created the bend by trapping the silt. Is this good? It probably is. Could it be done by anything else? Maybe. The native callistemons seem to do a similar job. Should willows simply be pulled out? Definitely not until we understand a lot more.

The road was straight for many kilometres but boredom was not a problem anywhere on the trip. The school kids often asked me where my iPod was but canned music, or any other recordings, would be an intrusion into the experience. By the time Ken picked me up and we loaded the kayak onto the ute the rain had stopped and it was getting dark. My Achilles tendon was sore but holding out. That night I was asleep by 7.00pm.

Chapter Seven

Dalby-Chinchilla

KEN DROPPED ME BACK ON THE ROAD before the school bus came through. Because people are friendly, chatty and interested in the strange kayak with the wheels, families came out to say hello and we soon had enough people for a party. I left Ken to keep up the social commitments while I covered as many kilometres as possible.

One of the farm dogs decided to come along. There was plenty to see, smell and pee on, so he was busy covering both sides of the road. No matter what I did he followed. I had it out with him at the first intersection, half an hour from his home. A lone man firmly instructing a dog in an otherwise empty landscape, "No, no! You stay." He dropped down by the road and watched me go. Looking back five minutes later, he was still on his stomach still watching. I presume he eventually went home.

The road was straight; not a bend in sight in either direction. The land was dead flat, except for the ring tanks. These are massive storages for water, with earth walls five metres high and each side hundreds of metres long. My route had taken me away from the highway. Although the Warrego Highway was the shortest route to Dalby we figured it was too dangerous. It was an extra 20 kilometres to go via St Ruth but far safer. There was also the added bonus of crossing Oakey Creek again to see how much water was still in it. There was none.

What had happened to the water? I was thinking about it again. Yesterday I had assumed that it was the aquifer recharging, water soaking through the ground into the underground reservoir. Looking at the ring tanks I began to wonder about pumping. Many of the farms have large channels, probably five metres wide by two metres deep, running along the lower boundaries to collect any water running off the property. These channels drain to a sump at the lowest corner of the property where huge pumps lift the water to the ring tank. They will often have another ring tank near the river so that if there is any water flowing past they can grab some of that too.

We spoke to a farmer about his giant travelling spray irrigator. It cost him a lot of money. The unit that we could see irrigates 376 acres (about 1.5 square kilometres). The sprayer is twice as efficient as the flood irrigation, halving his water use, but he continues to use flood irrigation on other parts of his property. With water being so precious he believes the cost of the sprayer was well worth the investment.

Despite having 50 millimetres of rain, and despite having a good flow of water from the hill around Toowoomba, Oakey Creek was bone dry a long way before Dalby. This didn't seem right. I began to appreciate first hand the environmental impact of water extraction and water capture.

After just 25 kilometres, Ken and Barb arrived. We loaded the kayak onto the roof and drove to the Dalby State School for a presentation. There were three classes combined, and Barb was involved in the discussion on sustainability. After the presentation one of the young girls asked me for my autograph. No one had ever done that before so I was quite flattered.

When we were unloading the kayak I put the rolled up laminated map of Australia down on the footpath. In the minute it took to unload the kayak and lift it onto its wheels, the map had gone. Ken headed off downwind looking for it while I took the kayak into the school. Despite five minutes of searching Ken came back empty handed. Later, I checked a bigger area but the map was gone. Jenny Cobbin had provided it for us and after a week I had lost it. I was cross with myself for losing something so thoughtfully donated. Besides, it was very useful.

I had been disappointed that I didn't have a camera with me on the river. Carol had lent me her expensive SLR with a couple of lenses but I couldn't take it on the kayak. Ken had his own camera and preferred to use that. With an hour to spare I went to the local camera store and purchased a waterproof and shockproof camera that I could wear around my neck. From that day until I reached the Southern Ocean that is where the camera was. Sometimes it was tucked under my jacket, sometimes it would hang free, but always it was within easy reach.

We had a presentation arranged at the caravan park where we were staying. Dick, the proprietor, plays the guitar and sings songs around a camp fire on Friday nights. Grey nomads, the retirees travelling around Australia during the cooler months of the year, love this sort of thing. As well as the sausage sizzle Dick puts on to make a night of it, the mayor was coming and it had been advertised in the local paper.

We had set up the screen and the sound system and had a mayor on the ticket so it was all reasonably professional. Not many of the retirees were worried about climate change, though. To be fair they will be long gone when the really serious impacts start to be felt.

One bloke in the audience had heard Peter Andrews talk a few years back. Peter is the author of *Back from the Brink*, a fascinating book about how we should look after our land. After hearing what Peter had to say, this fellow tried something new on part of his property. For decades he had sprayed weeds every year, reducing the weed patches to bare earth. The next time they appeared he slashed them without spraying. When they reappeared he slashed them again. After two years and no weed killer, he said that he had grass up to his knees and the weeds had stopped growing. He told his neighbours about this and some adopted the same method with good results. His brother has a similar property but is unconvinced. He doesn't want anything to do with that 'bullshit' and continues to spray every year.

The night was cold. There would be a frost in the morning and it got the locals talking. Universally they reckon on Anzac Day, 25th April, as being the time of the first frost for the year. For the past few years, though, it seemed to be getting later. This year the frost came seven weeks late. We asked if they thought this was due to climate change but they all said it was just a cycle. Things would eventually get back to normal.

A whole new coal mining industry is being established at Dalby. David Hood, the guy who had inspired us at the Ipswich forum, had almost convinced the council to set up Dalby as a model sustainable town. He didn't get it across the line though because coal mining would have no place in such an arrangement. Coal mining is expected to bring great prosperity and employment to the area.

The following morning it was only 15 kilometres out to the previous day's finish. Carol was coming out for the night as we were still only two hours drive from home. Before she arrived, I wanted to get 15 kilometres on the other side of town, giving me 30 kilometres for the day. Ken dropped me off and I arranged to drop in for lunch on my way through.

The first ten kilometres were fine. A couple stopped to talk about the Condamine. They had a property on the river. Being very religious they said "God bless" when they wished me well on my trip. They also thought that "God would be ashamed at what people have done to His rivers." I'm not religious in any way but I had to agree that any creator would be ashamed of the mess we'd made. Ken surprised me by driving back out six kilometres to say hello. After he left, I walked a few hundred metres and then my Achilles tendon really started to hurt. Being an ex-runner Ken had me stretching it every day and I also had a pressure bandage which really helped. This attack did not feel good. There was a sign post close by so I put my foot up against it and bent my leg forwards to stretch the tendon. It didn't make much difference so I decided just to put up with it.

This was the first day I had worn the proper walking boots purchased for the trip. Other days I had worn a cheap pair of joggers. Both were size nine. Ken had insisted that I needed expensive socks. They cost a hundred dollars for three pairs which made them the most expensive socks Carol and I had ever seen. Her comment was "Bloody hell, they had better be good!" They were thicker than other socks so both the joggers and the boots were slightly tighter than I was used to but still comfortable. The problem area on my tendon was right where the top of the boot rubbed. Perhaps it was the cause of the breakdown.

The next five kilometres were miserable. My leg was really sore and I limped badly. Overall I became very tired. It was a lot tougher than the day I had walked up the Toowoomba Range. I started to get overly optimistic and thought I had covered a lot more of the five kilometres into town than was actually the case. Up until then I was always close with my guesses of distances. What was happening with my mind? The footage of that day shows me complaining about being disappointed with myself.

Hobbling into the caravan park at last, I was ready to quit for the day. My leg needed ice and elevation for the rest of the afternoon. When Carol arrived my foot was up on a chair with a plastic bag of ice tied to it. My tendon was about the size of my thumb so I was concerned that it might delay the rest of the trip. By the evening it had started to feel better even though it didn't look too good. A slow walk over to the RSL club for dinner was just manageable. Carol is a nurse and we discussed the damage over dinner. The conclusion was that I'd tough it out unless it broke down completely, and hope that it would improve with time. I rang one of the physios in Brisbane the next day. She said that what I was doing was fine and that there was nothing more that could be done.

🛶

The next morning Ken and I went to the office to interview Dick on camera. Before purchasing the caravan park he had owned farms in the Hunter Valley. Dick despairs at what he perceives we have done to our rivers. As a property owner he had seen his share of droughts and floods. "Water is for the rivers," he says. "It should run free. But a few get it all." Dick has no doubts on who is to blame. "The blokes who are supposed to look after allocations and police pumping and collecting water do not do their jobs properly. This is unfair on every Australian," he laments. He firmly believes that under most circumstances the water just does not get into the river system.

CHAPTER SEVEN

Dick had only been in the area for four years but it had been on water restrictions all of this time. "Most people are pretty good," he reckons, "but there are some who don't care. They come in and wash their cars and their vans and don't care about anyone else." When it comes to water Dick is well and truly a socialist. "Water is life and it is owned by everyone," is his philosophy.

For the first few kilometres west of Dalby it was necessary to follow the Warrego Highway. Luckily the verges are very wide so there was not too much danger. Nonetheless it was a relief to turn off on the road to Kogan. It was another dead straight road. Surprisingly this is never dull or boring. You slow down, enjoy the moment and your frame of reference changes. I was becoming expert at picking something in the distance and guessing how long it would take to get there. Before long the others arrived and we had morning tea. Carol walked with me for ten kilometres but she hadn't brought the right shoes and socks and started to get blisters. She took the car and we arranged to meet at the bridge over the Condamine River for lunch. My leg was a bit sore but nothing like it had been the day before.

As I approached the bridge a photographer for Brisbane's *Courier Mail* arrived. It amazed me how many shots he took. It had to be something in excess of 500. He wanted me down in the dry river bed with the kayak. I thought it was stupid because readers might think I was planning to drag the kayak all the way along river beds. Ken was happy with the idea, so down the bank it was, posing for another squillion photos.

The river bottom was sandy and had trail bike tracks on it. There was no doubt it had been a long time since it had seen water. Seven months later there was heavy rain and the river was quite high at Warwick. I went out to have a look and saw the area covered in muddy water. Staggeringly though, there was no flow. The river had come up, the water had passed, and that was it. It was all gone. This is not a river as I understand a river. It is a drain, just like a stormwater drain, but not made out of concrete.

I travelled on until late afternoon when we loaded up and headed for Greg and Tammy's. Greg is a mate who is managing director of Aquatec and has been with the company his whole working life. He and Tammy spend at least every second weekend on the family farm at Chinchilla. Over the years we had heard a lot about the place but had never visited it. Carol and I were both keen to see it. There was an added bonus in that Greg provides superb red wine. He has always cellared good quality wine as long as I can remember. We reckon keeping a few bottles from Christmas to New Year is about the limit of our cellaring ability. In other words we are hopeless at keeping it but pretty good at quaffing it. Conveniently, our skills are most complementary to Greg's.

When we drove into the farm I was with Carol in the Astra. Ken and Barb were behind. We were on a rough track and due to the recent rains there was a stretch of soft mud. Carol drove into it before realizing what she had done. Luckily she has done plenty of mud driving and just kept the revs up and the car slithered through. Ken followed with the ute and trailer.

Even though I hadn't been out here, I knew they had a permanent water hole because about 20 years ago I gave Greg an old wave ski for the kids to paddle. Both Greg and Tammy are passionate about the farm, especially the creek. Times had changed and I asked Tammy to briefly summarise their situation.

> We have a grazing property bounded by Charley's Creek, (a tributary of the Condamine River) and also with a natural lagoon system through the length of the farm.
> Before the issuing of irrigation licences, our property had a really reliable natural water supply, having never been 'out of water' in recorded history. It just didn't happen! We have now experienced this twice in recent years – with dire consequences, and many more worrying times when the creek levels have dropped dangerously low. The water extracted by irrigators and water harvesters is a point of much contention in the community, especially in dry times, as non-irrigators struggle to access enough water for basic stock and domestic needs and the creek itself continues to degrade.
> Though the state government purports to be working to achieve the sustainable management of water with the implementation of water plans, in practice, little or nothing is being done and the issue of over-

allocation is being completely ignored. The impact of the existing irrigation licences has not even been considered in developing the new plans, and even individual water harvesting licences which allow extraction of up to six-sevenths of the flow from the stream upstream of our property remain unaltered. I believe that the general population would be absolutely horrified if they knew what is really allowed to happen. I don't know how anyone could justify such environmental vandalism, but our Queensland Government seems very comfortable in the knowledge.

Our creek is relatively small and its flows are erratic. It exists for most of the time only as a string of waterholes – a fragile and vulnerable system. Licences now exist, however, which allow extraction from these waterholes with no limit on the level of extraction. It really irritates me to hear politicians asserting that there's no over-allocation in Queensland, while licenced water extraction of six-sevenths of the flow, and then of a further three quarters of the remaining flow occurs downstream (and that's only two of the licences in our catchment). I've been told by a departmental officer – 'we can't touch the licences or we'd have World War III'.

In addition to this, considerable water never reaches the creeks and rivers any more due to the activities of 'overland flow harvesters'. So the ongoing degradation of our creeks and rivers will continue, at least until enough people become aware of the real facts, care enough to change things and demand change. I concur with Tim Flannery and John Doyle (from 'Two Men in a Tinnie') when they stated that the loudest voice is the voice of irrigators, because the voice of irrigators is the voice of money. First and foremost must be the voice of the environment. History has shown that civilisations that have ruined their river systems have ceased to exist!

Tammy's words are somewhat sanitised and do not include the bastardry that goes on. It all fits the pattern that Dick and others are saying. "Water is money and a few people get a lot of it." Someone is eventually going to get hurt and maybe even killed over water.

Despite Tammy growing up in the area and despite Greg being heavily involved in the farm many locals consider them city slickers. Upstream of them the creek has a dam on it. Below the dam the farmer, who is over

70, has no water and hand-feeds his cattle. By law the dam must have a 100 millimetres valve in it that is to be open unless water is running over the top. This valve was broken for years until concerted efforts forced the owner to fix it. When it was fixed he refused to open it.

> Greg said to the guy with no water "Why don't you do something about it?"
> The response was "He is my neighbour".
> "Well I'm your neighbour too and don't you come and tell me every time I need to fix a fence?"
> "Yes, of course," said the old farmer, "but you're a city boy."
> "Don't I stand by you and fight for you at the water meetings? Aren't I your mate? He certainly is not your mate," replied Greg.
> "He's my neighbour."

That was the crux of it. On the one hand there are gentlemen working to ancient codes of conduct. On the other there are people willing to take whatever water they can get and to fight tooth and nail for it. Administering this mess is an incompetent state government department. There are landholders pleading for common-sense regulations and then there are those landholders who have just given up even trying to reason with a state government that fosters obscene water extraction.

In the morning we had a look around the farm and saw a giant jacaranda tree that Greg had told us about in 1985. For the past 30 years he and Tammy have kept rainfall records here at Chinchilla and also at Karalee where they are our neighbours. Karalee is near Ipswich and is 20 kilometres in a straight line south west of the centre of Brisbane. With minor discrepancies, the rainfall in Karalee has been much the same as at the farm at Chinchilla. When I first heard this I was astounded. Brisbane is not right on the coast and Karalee is even further west, but even so I had always pictured the country two hours' drive west of the Great Dividing Range to be comparatively dry. Greg's rainfall readings made it clear why the Darling Downs is such a prized agricultural area.

Much as we had enjoyed the farm, it was time to hit the road and get back out to Kogan where I could resume walking. Again I travelled with

Chapter Seven

Carol. She was going back home after dropping me off and I wouldn't see her again until 5th July when Ken and Barb were taking five or six days off to go to Sydney for a wedding. That was three and a half weeks away. To get to Kogan we had to cross the Condamine River and were surprised to find water in it. "Bugger going down to Kogan," I said. "Let's use this water while it is here."

The area around Kogan is ridge country. It is only about 20 kilometres in each direction until black soil is dominant and it becomes farm country. There had been just over 100 millimetres of rain in the past three days so even this small catchment, because it had no development, was contributing water to the Chinchilla weir. This seemed to confirm the observations of Tammy and Bill that it is the extractions for irrigation that cause the river to dry up.

As we unloaded the kayak a car stopped. Two local farmers clambered out to check us out. They'd heard that a 'greenie bastard' was coming through and giving farmers a hard time. I explained that we were here to listen and learn. Ken said he'd love to set the camera up and record their comments. At first they were distrustful and reluctant but I said that they were free to say whatever they liked and could tell me if they didn't like my questions. We promised only to use the footage if they were happy.

They had received a fair amount of abuse and were sensitive to criticism. We started with farm efficiency. Greg had told me farms in the area had increased efficiency by a factor of four in the past 30 years. These guys agreed but said that hadn't translated into more profits. Farm costs had gone up many times over but revenue had not. As an example, sorghum has only doubled in price in 35 years whereas costs have risen as much as ten times. It took quite some time to tease out what they believed. In a nutshell they agreed that the river was over allocated. But what were farmers to do? They had to make a living. They only used the water that they had a licence to use. The blame lay squarely with the government which had no idea of what it had done and what it was continuing to do.

Carol and I said goodbye and I paddled away towards Chinchilla. It was easy paddling, just water with no trees or logs. After less than an hour I was on the weir pool of the small dam for the town. The water was very turbid but there were half a dozen groups of people fishing, spread over about 10 kilometres. The water level seemed to be about three metres lower than normal. I judged that, because the steps on the jetty were three metres above me. There was a lot of algae on the water in some places so I took photos for Amanda to identify. If someone doing a doctorate on blue green algae couldn't tell what it was then no one could.

The weir pond was full of Noogoora Burr. It is a dead looking weed with burrs that latch onto anything that goes past. A week later you can still be picking burrs out of socks or kayak ropes. Noogoora Burr is throughout the Murray Darling system. Nasty as it was, it was just a foretaste of all the prickly things I'd encounter before I reached the sea.

By jumping in the river and paddling to Chinchilla I had cut 15 kilometres off the trip that I'd have to make up. The aim of the trip is to travel from Brisbane to Adelaide in a continuous line and this decision had broken that line. The shock of realising that the water only stayed in the river for a few hours had made me decide in favour of using the water while it was there. I would make up the 15 kilometres tomorrow.

Ken and Barb had been down to the Kogan pub to interview Tony, the publican, while I paddled. I was disappointed that I had missed that. Tony is a character and had plenty to say to Ken and the camera. Many people expressed serious doubts about cotton farming but few would say anything publicly. Tony had the guts to tell it as he saw it.

The people from the Kogan area are known by some of the farm people and the Chinchilla people as ferals. Tony loves the area though. He has been there 15 years and prefers it to the more productive land. "At Kogan," he says, "There is diverse wildlife and there are frogs. There are no frogs on the Condamine. It has been raped and pillaged by cotton farmers and people who are not managing it properly. There are plenty of stories of the river running backwards to pipes big enough for you to crawl up. The water is not shared. Millionaires make extra millions using a river system that belongs to everyone. The river system is in crisis." Tony drives to Dalby at least once per week and keeps his windows closed when the crop dusters are out to avoid breathing the clouds of poisonous dust.

I left from the caravan park about an hour after the sun had come up. Walking to the weir would make up the extra kilometres I had skipped by jumping onto the river the previous day. The day was crisp and clear with a light frost and, again, I could watch people go to work. A truck full of council workers waved. The area seemed friendly, like Toowoomba.

Chinchilla has irrigation which brings wealth, it is near gas fields which bring wealth and the new coal mines were about to bring even more wealth. The main street is beautiful with large trees and a really comfortable, country feel.

As usual, I walked towards the traffic. On the road out I did my customary wave to the first car. That didn't go down well. They gave me the finger and mouthed that I should get off the road. I continued to smile and wave but after they went past I thought 'Go back to bed you grumpy prick.' The next car was the same. And the next, and the next. 'What is going on here?' I thought. Soon I came to recognise what I call the Chinchilla salute. It is the middle finger extended sideways in a gesture to get off the road. Usually it is accompanied by angry words mouthed through the windscreen glass. I can lip read "Get off the fucken road," fairly easily.

Talking to a local councillor a few days later I learned that many people think Chinchilla is an unhappy town. "Funny lot," she said. "Maybe too much like a city with all the subdivisions." It seems that 'progress' has just made people miserable. Tammy went to school there more years ago than she wants to say, but she remembers it as a happy town.

I can respond with the same level of maturity as anyone so when I turned the corner towards Condamine I thought, 'Fuck you too.'

Chapter Eight

Condamine – Surat

FIVE KILOMETRES BEFORE THE TURN, I had an interview with the ABC country program. While the interview was finishing the interviewer asked me how I was getting on with my phone.

"How is your Telstra coverage?"
"Well," I responded, "in places it is fine, but I'd have to say the Next G coverage maps on the internet are optimistic."
"Thank you Steve, and good luck with the rest of your trip," she said.

Ken and Barb arrived for a surprise visit. They brought an apple turnover for morning tea which was most welcome. We discussed the 'friendly' reception from most drivers.

A bloke pulled up to say hello. He was working in the gas fields about a day's walk away.

"Heard you on the radio," he said.
"How did I sound?"
"Talking about Telstra you were. Good on yer. The next G coverage is nowhere near as good as they reckon."

I thought it was interesting that out of the whole interview he had only picked up on a brief comment on Telstra.

Turning right towards Condamine half an hour later a van roared up beside me.

"You're on the radio!" yelled the driver.
"Yes, I know. I was on an hour ago."
"No mate you're on now – listen".

He turned the radio up and opened the door. The encounter made both our days. We listened to the last bit of the full interview and he turned round to continue on his way. There was a valuable lesson in the whole experience for me. Any interview you do can be cut and pasted into whatever context the broadcaster wants. My last words had been quickly slotted into an interview with a Telstra manager; the bulk of the interview was played later.

Ken arrived at 5.00pm and we loaded the kayak onto the roof for the drive back to Chinchilla. "Well done," he said, "38 kilometres today. Obviously you just need someone to get you upset and you cover more ground." Maybe so but I was still disappointed. I was still short of my goal to travel 40 kilometres per day. We knew exactly how far I had come because Ken had decided to put the GPS on the kayak. The speedo in the ute could tell us as well but it was much smarter to log the trip on the GPS. This would be especially important when paddling along the rivers.

Pain management was becoming an issue on the trip. Certainly, from the date of the accident pain had been a part of my life. I mostly dealt with it by putting it to one side and trying to ignore it. Doctors and nurses ask you what level of pain you're at. One is low and ten is the highest you can take. It is a very individual judgement. One person's ten can be another's six. Six is the level that I can ignore by thinking about other things. Seven is something that really starts to bother me and ten is like the night in the Alice Springs Hospital when I was a smarty pants and decided to cut the morphine out. I had woken up feeling like a shark had me by the ribs and was on the buzzer quick as a flash, requesting morphine from the nurse. Davo caused a ten when he shoved his arm up my jacket on the Plenty Highway. Eleven doesn't matter because it's more than you can take. After ten I black out.

Throughout the day the pain level was six when walking. Putting the weight back on my feet after a rest it would rise to seven. When Ken took me back to a caravan park for the night it was a struggle to walk after getting out of the car. This would quickly improve though and after a beer or two all would be fine. During the night when I went for a pee I had to be careful not to fall over. This was due to lack of lateral strength between the bones along my feet. It was becoming clear that managing the physical aspects of this trip was all about recovery. I finished the day with nothing left in the tank, recovered overnight and then went through the complete cycle again the next day.

Condamine is only a day and a half from Chinchilla so there was plenty of time to get there by lunch and then go over to the school at 2.30pm. The system of visiting the schools was working extremely well. Jenny Cobbin would contact the school by email and phone about a week before our arrival. Barb would then talk to them and organise a time close to when we would arrive. It was not possible to have a definite timetable until a day or two before getting there so we accepted that presentations in towns might not be viable.

With a large part of the Darling Downs behind us we were beginning to form some impressions about the land and the people. Apart from a significant number in Chinchilla, the people had been overwhelmingly supportive and friendly. There was no doubt that all farmers are very efficient users of water. Some had lots of water but some had none and this was probably the cause of the deep divisions in the community. Having seen Toowoomba split more severely by a recycled water debate than any Australian community by any debate on religion or politics, I was not surprised at the level of passion. Water can certainly polarise.

Ken and Barb had gone on ahead to set up camp in Condamine. They came back to provide a very welcome morning tea and rest in the fold up chair. Ken asked me to call him on the radio when I could see the town sign. When I did, he came out to film me arriving into town and then went back to the tent to film the entry into the caravan park. A few minutes after he left, bang, it happened again. My Achilles tendon went into severe protest mode. The radio didn't do any good because Ken was already out of the vehicle all set for filming from in front of the tent. My mobile phone was flat so there was no way of making contact.

It was less than a kilometre to go but it would have been nice to stop right there and get some ice on it. Rather than push on hard I decided to take it really slowly so it took twenty minutes to go the few hundred metres to the caravan park. Eventually Barb came out to the gate and I waved but she was unaware of the trouble I was in. Making it to the tent I flopped down in a chair and lifted my leg up.

"You need to get ice on that," said Barb.
"I know," I replied, "but right now it needs a rest before I can go and get some."

The pain subsided rapidly with the foot up. It was only an hour until we were due to speak at the school and I was too comfortable to worry about getting any ice. Besides, there was some water in the river here, so tomorrow would be a paddling day.

Rather than lift the kayak onto the roof I walked over to the school which was only about 400 metres away. My Achilles problem was no longer acute. It is amazing what rest can do. As usual the kids were fascinated with the kayak and the stories. This school was to be extra lucky though. Ken announced that all being well, the next morning the Channel 7 helicopter would land on their oval and they could meet high profile weatherman, John Schluter. Wow, a wheeled kayak towed by a man in a funny blue hat one day, then a famous television personality in a helicopter the next. John was coming out to do an interview as part of the channel's environmental section in the news.

Ken was still alive the next morning despite wearing a blue jumper deep in maroon country during a rugby league State of Origin game. All he had to do to survive that night was to sit quietly and watch Queensland beat his beloved New South Wales. I suggested counselling might help as he would need to get used to this pattern but my advice was not enthusiastically received.

The helicopter landed. John Schluter spoke to the kids and they were all allowed to file through the cockpit. The cameraman filmed me walking with the kayak and then we set up down at the river. John and I chatted for a while as he was learning about the trip and who I was. He later confessed to Barb that he needed to find out if I was a nutter. He rapidly established that we had a very serious message and were all well within our senses. John said the view from above was of lots of water lying on the ground. With falls of up to 150 millimetres in some areas I was not surprised. It only strengthened the widely supported view that the river channel itself should have been flowing strongly.

The interview went well. When it was finally aired it went for more than two minutes. The producer told us that this was a long segment for a news item. It did seem like a lot of money for such a short time, involving three people plus the cost of the helicopter for more than half a day, not to mention the greenhouse gas implications of a helicopter versus a much more economical car.

During the interview a couple of older women came down to watch proceedings. One told us about the river when she was a girl. She pointed to the old boat in the park and told us that was how people crossed the river then. It was deep and clear with reeds along the sides. Kids swam in it and drank from it. Even 20 years ago people would fish by dragging spinners through the water. Today no one would drink it. You wouldn't need a boat because you can walk across it. No fish would swallow a spinner because it would not be able to see it in the murk. "You tell them about our river," she said, "because I don't want our children to have to live with it like this."

To complement the talking heads, the news crew needed some paddling shots. This involved countless trips up and down a fifty metre section of river before they finally sent me on my way so they could get some aerial shots. This was thrilling. There were a couple of vertical drops of about a metre. The helicopter was hovering in front, blowing wind gusts along the water. The noise and the wind stirred up a lot of adrenalin and I concentrated on looking natural but watching the helicopter

CHAPTER EIGHT

out of the corner of my eye. Any thoughts of fuel and greenhouse gases were washed away by the excitement. And then they were gone, soaring high into the sky and away towards Brisbane.

My job was to keep paddling but it got tough in many places so I made very slow progress. There was simply not enough water to float the kayak above what were quite sharp rocks. Ken and Barb found a crossing 15 kilometres downstream and settled down to wait. It was not until we were four kilometres apart that the radio worked. They were most relieved to make contact after waiting for three hours, not being sure they were on the right river or whether I was on the right anabranch. We each had a 1:250,000 topographic map and a GPS. Ken had plenty of time to kill so had learned how to input co-ordinates on the old GPS. With my co-ordinates communicated to him by radio, Ken could establish where we were in relation to each other. It took another hour to reach them. They were at a causeway and had come in from the northern side.

My belief was that to get back to Condamine, the quickest way would be along the southern bank but we went the way Ken knew. This took over half an hour. The route took us past a huge cattle feedlot. The scale of the operation was staggering. There were feed trucks lined up as well as empty cattle trucks. There were great mounds of cattle shit – presumably stored for fertiliser. All the paddocks were bare due to the drought. It was surreal, it stank and it was offensive.

The scratches on the kayak were deep, Adelaide was a long way off and it needed to be preserved. With only 20 kilometres paddled for a day that finished close to dusk, I had to go back to walking the next day. The helicopter pilot had announced that Adelaide was only 507 nautical miles in a straight line, which is about one thousand kilometres. Unfortunately for us there were a few bends on the way and we were headed south of Adelaide to the Murray mouth and then back up.

There was a lot of noise on the oval that night and it kept up way past my bed time of 8.00pm. It sounded like rugby training but I couldn't be sure. The temperature was about five degrees but they sounded like they were having a great time. What they were doing didn't become evident until we reached St George.

Terry Loos called to tell us about his wife's sister, Pat Stallman. She and her husband, Bil with one 'l', had a property downstream of Condamine. Terry is an old AWA mate. He's AWA's policy officer, so is a very valuable resource. The fact that he had relatives along the trip was a bonus. Some years ago Terry had told me about his brother-in-law. "Crazy bastard," he noted. "Just like you. Tears around the place on a motorbike like a teenager." More importantly for us, they had purchased the property between them and the river. This came with a comfortable house and it was only used by their children who were not there at the moment. Peter and Jenny were travelling out to see us this weekend so the timing was perfect. We could all stay in the house. Peter was the bloke who sold Watergates for us. Jenny is his partner. Peter is an excellent cook and he was to supply dinner for Friday and Saturday nights. Barb's cooking was very good but it was camp cooking done under circumstances not conducive to culinary storms. I knew Peter would be spending a lot of time preparing for this sojourn and could hardly wait.

Over breakfast we looked at the maps. The Condamine-Meandarra Road is bitumen and runs parallel with the river about three kilometres from the southern bank. The crossing that I finished at was not shown. Ken wanted to take me along the bitumen and drop me off parallel with where he estimated the causeway was. I argued that as there was a causeway and a gravel road that it must meet up with the bitumen, so we could drive down the bitumen and then find the road.

> "No!" Ken insisted. "I will take you parallel with the causeway."
> "I need to go back to where you picked me up."
> "It makes no difference to you. I will drop you on the bitumen."
> "You will take me to where you picked me up."

Eyes fixed firmly on mine and close to my face Ken hissed "Stephen, I will drop you on the bitumen."

Never back off. Never give ground. These are my rules. I stared back, moved slightly forward and said "Ken, I will get back to where I stopped yesterday – somehow."

Ken stared, eyes wide. I held my ground. He finally dropped his eyes. "You'd be fucked without us," he said.

We went back the long way round, via the stinking feedlot. Our opinion of it wasn't any better than yesterday's. Barb stayed at the camp. Ken was back to normal. "That must be one of the dummy spits Barb talked to me about," I thought.

Walking up from the causeway the landscape was like something you would expect on the moon. There were washed out gullies, craggy rocks and bare dirt. Despite the rain there was zero vegetation. Ken passed me on his way to the bitumen, less than two kilometres away as it turned out. We agreed to meet on Sunnyside Road, the major gravel road on the south side of the river, in time to get to the Glenmorgan school before they went home. At midday I was to call on the satellite phone and confirm my position.

That morning they had two interviews scheduled. Bill, the local State Emergency Services controller, said that the last flood had been in 1996, eleven years ago. Now the normal floods were not coming. "Experts say that the ring tanks have no effect," he said. "If they stop the overland flow into the river that can't be true." He was more upbeat about the fishing, having heard of two cod being caught and released in a water hole downstream just two weeks ago. "They are learning to feed off the carp which must be a good thing," he opined to Ken's camera. On the water extractions he was optimistic. "You used to get a licence based on how many acres you could irrigate. Now it will be on volume and it will be metered."

It is astounding that such an obvious and logical step in managing water should be new in 2007. Back in Chinchilla the area criterion works to the advantage of some melon farmers. They only have to count the area under the plastic that they put down to cover the rows. If the melon vines happen to expand outside this, that's not their fault. It would certainly be unfair to ask them to count the whole paddock as their licensed area wouldn't it?

The second interview was with Elizabeth who had implored us the day before to tell the story about her river for the sake of her children.

When she was a schoolgirl the teacher would take the kids to a sandy part of the river once a week for a swim. Then you could see the bottom. Elizabeth says if the river is always muddy now "there's obviously erosion coming from somewhere." Like Tony back at the Kogan pub, she talked about the river running backwards when the pumps are on. "The water harvesting is abused. It is the big guys that win, average landowners do not irrigate." On two occasions Elizabeth lamented, "The river is dying."

It seems that if you do not make money by taking water from the river then you're probably hostile to its degradation. If you do make money from it, you justify to yourself that it is all worthwhile and the damage you cause is not as bad as everyone makes out.

After Ken passed me on his way back to Condamine, a woman from the nearby farmhouse came out. Everyone has an opinion on why the river is in such poor condition. She said, "The carp eat the weeds that clean the river." We were to hear this many times. What is the truth? Folklore in the bush is very strong and it may or may not be based on fact. After all, one strong message on the bush telegraph was that a greenie was travelling down the river to try to tell them how to run their farms.

Walking along the gravel road was slower and harder than on the bitumen. The kayak felt heavier. It was hard work pulling it over cattle grids. The wheels were just big enough that they didn't slip all the way down between the bars. Luckily the front wheel was already off the grid before the back wheels came onto it. This meant it was possible to walk across the bars only having to deal with the front wheel. When it was clear, it was time to apply full strength and speed, bouncing the back wheels across. After the crossing a good check back was required to ensure that the violent juddering hadn't shaken something off the kayak.

Some way down Sunnyside Road a bloke drove up from behind. "Heard about you," he said. "You want some lunch and a cup of tea? How about a bed for the night?" His place was 11 kilometres back along the track so despite this most generous offer I thanked him and stuck with my peanut butter sandwiches, explaining we were staying at the Stallmans' place that night. He knew them and showed me where their property was on the other side of the river. His comment on our goal of reaching Glenmorgan by 2.30pm was, "Well you won't walk there by then, it is 40 kilometres."

A young couple stopped and offered me lunch as well. From their place they reckoned that I could paddle down to the Glenmorgan weir which was three kilometres downstream. It was more great hospitality but there was not much point wandering off the road for just three kilometres on the river.

At noon it was time to call Ken on the satellite phone. There was no bloody phone in the kayak. Ken must have had it out to learn about it. This was no big deal. Sunnyside Road was on the map so all they had to do was drive along it until they found me.

The river was close to the road, the trees along its banks marking its course. It seemed that where we would be staying that night was across the river and back upstream a few kilometres. Ken and Barb arrived at 2.00pm and we loaded the kayak onto the trailer and hastened to Glenmorgan school. Barb and Jenny had made the arrangements as usual so the students were expecting us.

It was a Friday afternoon and there was a religious instructor talking to the children. We were concerned that there wouldn't be enough time after that for us to talk to the children before the school bus arrived. It transpired that the woman we were talking to was the bus driver and she was very interested in the trip and the kayak so we needn't have worried. Nonetheless we were conscious of the time and slipped through everything quickly. Being Friday, a few of the people there were to meet at the pub later and naturally invited us. Unfortunately we couldn't go. Peter and Jenny were arriving to meet us and then travel over to the Stallmans for the night.

We found the Stallmans' property and then Bil found us. He was on a quad bike and invited us to follow him to the house. He took off quite fast and it was all we could do to follow his dust trail. There was no way of keeping up. Terry's description of his mad brother-in-law came back. At his house, Bil got into a 4WD and we followed him slightly more sedately to our accommodation. It was a terrific night, great company, good wine and an excellent seafood curry dished up by Peter.

Bil is a character. That's how he spells it. "Bil with one 'l'," he says. We had interviewed Bill near Toowoomba; there was Uncle Bill at Oakey; the local State Emergency Services bloke was Bill. Maybe Bil with one 'l' made sense. "Tried cotton," he declared. "Too hard. It owns you. Lots of monitoring and when the cotton tells you that something is needed you have to do it straight away. I reckon it gives you ulcers. Anyway, I made a few dollars out of the crop and decided that was enough." With the two properties the Stallmans now have 20,000 acres. The original property is 15,000 acres. The new one takes them to the Condamine River. Bil hadn't seen the river until he bought the second property even though he had lived there for 20 years!

Bil was about to plant oats. Greg had planted oats back at Chinchilla. It seemed just about everyone was planting oats. Bil's would be sold to the local feedlot. Greg's were for their own cattle. The equipment Bil uses is 15 metres wide. The tractor is an enormous air conditioned beast. All you do is turn it around at the end of a row and the GPS controller does the rest while the operator watches a video or reads a book. When I was on the road two days later Ken and Barb were shown how it all worked and were allowed to have a drive.

There was another good frost that night. Morning temperatures had been around freezing since the rain stopped the other side of Dalby. All the blokes staying at the farm came over to the river where we unloaded the kayak and I set off. We were not exactly sure where we were on the map but knowing the weir was three kilometres downstream from where I was yesterday helped us to figure out the relative locations.

My Achilles tendon was sore again, probably from the cattle grids but maybe from pulling on rough roads. It was a welcome relief to be paddling. The paddling was easy with plenty of water. There was certainly a weir somewhere ahead. The trees were teeming with birds, all screeching to let the world know what a perfect morning it was. This was like the cold, crisp mornings at home, paddling to work along the Brisbane River, except it was a long way from home and on a strange river.

By 10.00am the wind had come up and added the rustling of leaves to the other bush sounds. At 11.00am the rustling became a roar. The trees didn't seem to be moving much despite the extra noise. There was a grand old homestead on the bend. Hang on, what did Ken say? "Listen for the weir." Next thought? 'Bloody hell, there it is 10 metres away.' Pulling into the bank I surveyed the scene. It was amazing that at a distance of 10 metres you can be oblivious to a two metre drop right in front of you. Luckily the bank below the weir was partially concreted so it was possible to pull the kayak out and easily put it back in downstream from the weir.

From then on the day was spent fighting one obstacle after another. My shirt was ripped on a barbed wire fence. The front of the kayak got wedged under a log that was tapering down to the water. To get the kayak out I needed to move the back around to the left. It was hard up against another log in this direction and there was no budging it. Water was rushing underneath. 'Shit! Here we go.' Maybe stubbornness prevailed. Whatever it was the kayak squeezed through an impossible gap leaving me with a few bruises and a bit less skin.

Peter and Jenny experienced the waiting game that Ken and Barb copped two days earlier. We had agreed to meet at the road crossing at 1.00pm. Peter is always early so he arrived at midday. I turned up at 3.00pm after making radio contact at 2.30pm. Ken and Barb had come out to the crossing as well but were smart enough not to get there early.

It had been 20 kilometres to the weir and the total for the day was only 36 kilometres. We loaded the kayak on the roof and went back to the Stallmans where Ken and I prepared a big fire a hundred metres from the house. Bil came over to see if we were OK. Or maybe he came over for a beer and a glass of red. It was good to see him again. Peter prepared a magnificent feast of lamb shanks and vegetables.

The next morning Peter dropped me back at the crossing to walk towards Glenmorgan. The water level had dropped and paddling was impossible. He returned to get Jenny and his vehicle for the drive back to Brisbane. Ken and Barb hooked the trailer back onto the ute and went to play with Bil's big toys. There were some cattle grids with extra wide openings on this leg and they caused some delays. 'Bloody farmers, they put barbed wire across creeks and now they set traps for kayak wheels,' I thought. To be fair though, probably the only wheeled kayaks around were either with me or on the trailer.

My right index finger was sore and there was pus coming out of it so I had obviously done something to it on a fence, a log or something else. My Achilles tendon was sore. It looked like whether paddling or walking, I was going to be injured.

We had planned to be at Surat to meet the mayor on Sunday and were behind schedule due to the difficulties in the river. Ken and Barb picked me up at 11.00am and we went to meet her. There was a barbeque in the park with the local fishing club and the ever present grey nomads. The park was on the river filled with water backed up from the weir. Donna, the mayor, made a welcoming speech, spoke with pride about their section of the river and apologised that there was no water down to the Murray. Ken interviewed the local fish stocking club guys who think the river is looking a lot healthier over the last couple of years. They pointed out some small she-oaks and sedges, which are small reeds, that were starting to establish themselves down on the banks.

"What do you put it down to?" said Ken as he filmed them.
"Probably the reduction in the carp," was one answer.
"Haven't had any big floods, and that would have helped."
"Would have helped how?" enquired Ken.
"No water off the Dalby flats."
"What does that mean?" from Ken drew an apprehensive laugh.
"I can cut it if you like," said Ken.
More nervous laughter. "You'd better cut it."

These guys thought they were starting to win. The damage they had seen from cotton growing was declining. Runoff carrying heavy chemical loads was less of a problem. They didn't want to tread on any toes and set things back by saying, particularly on film, that they were pleased not to have had chemical laden water from the farms around Dalby.

After a very convivial lunch we drove back to the road leading into Glenmorgan where I started walking again. Ken and Barb then drove into town to set up camp. Late in the afternoon, about two kilometres from town a local bloke stopped for a chat. His stories about the river and his farming methods were fascinating.

"Fancy a beer?" I asked. Although he had been to the golf club and was on his way home he agreed to meet in the pub. I set off at a good clip, grabbing Ken as I went past, insisting that he and Barb needed to meet this guy. You don't have to twist Ken's arm all that hard to get him to have a beer.

The pub was an interesting experience but we didn't get the information we wanted. There was another bloke there who dominated the conversation. Ignorant conversation it was too. The man was a smart arse bully. When we shook hands he grabbed my hand and squeezed it as hard as he could for as long as he could. It didn't hurt and I gave as much as I got but all I could think was 'What a prick.' With the pub interview shot to pieces Ken arranged to meet Greg Hoadley, the bloke I had met on the road, at their farm the following morning.

The pub has a side trade in pizzas. A young couple with a property out on the ridge country were having a drink while they waited to take the pizza home. They offered us a piece before they left; such is the generosity of many of these people. We had been talking about water and soil erosion. They said that when it rains the gullies just get bigger. After a storm they go out to see how many trees have fallen down with their roots washed out. They felt this was tragic but what could they do? Their story made me think that some people shouldn't have charge of land.

The next day was a good start. It is always good to start or finish the day at a camp site and avoid driving to or from camp. Ken picked me up 38.56 kilometres later, still short of my desired 40 kilometres a day.

They had set up camp at Surat after interviewing Greg Hoadley. Obviously Ken had run into some problems downloading film after setting up the camp. "It's alright for you," he said. "All you have to do is walk. I have to worry about all the hard bits."

I kept my mouth shut, struggling to believe what I had just heard. What is the response to something like that? Any reply is going to be inflammatory or condescending.

The interview with Greg Hoadley challenged conventional wisdom but it makes sense. All his life Greg had sought answers to why good grass cover would disappear. He had tried running many fewer cattle because conventional wisdom was that too many cattle were the problem. All this did was put pressure on his viability. In desperation he tried locking up part of his land but the kangaroo and wallaby populations exploded. When they bought the property, located on the junction of the Condamine and Balonne Rivers, three years ago they noticed that there was a lot of erosion in the area and a lot of silt in the rivers. This was a good place to trial new ways of doing things.

They now use a method known as inside/outside management or 'cell pastures', which is another way of saying holistic farming. Although the real explanation is far more complex, he puts a lot of cattle onto an area to 'flog it down'. Then he pulls the cattle off until he sees that all grasses, especially the ones he wants, have recovered. The cattle are then put back to repeat the process.

Why could this work? Cattle eat what they like to eat. That means they will take out their favourite food and when it is gone move onto their next favourite. This will then selectively remove the things they like and allow the things they don't like to gain traction. If they are grazed heavily, they have to eat the lot. There is no favouritism. If the cattle are taken off in time, all species have an equal chance to recover. The trick is to leave the cattle long enough to eat it down but not long enough to destroy it. The theory sounds good and Greg reckons it is working for him.

When Greg was in a bar in Nui Dat during the Vietnam war, a mate banged an unexploded ordinance on his own knee. Greg was too close.

It blew his left hand off and most of his teeth out. "Shit," I said, "it was lucky that didn't kill you."

"I have always been lucky." he said. I love his attitude and tell that story as an inspiration to anyone complaining about their luck.

More delving into Greg's philosophy uncovers an astounding depth. Diesel is expensive. It also produces carbon dioxide and is environmentally unfriendly. Instead of using machines, Greg uses the impact of his animals to beat down the unwanted growth, 'just as happens on the great plains of Africa.' His approach also allows more organic matter to build up in the soil, burying carbon and increasing its ability to absorb and hold water. Although records indicate that rainfall here should be 575 millimetres per year, Greg thinks he can get the property to be sustainable with only 350 millimetres per year. Because of the sequestered carbon, he's doing much more than his share in the battle against global warming. This method of pasture management even has spinoffs for his aching body. When the cattle have grazed a paddock, when they have trampled the remainder of the vegetation into the ground and when they have spread their dung around the whole paddock, all he has to do is to open the gate, blow the horn and the cattle move to where he wants them. The hard work has all been done building the fences.

Greg has many other observations. He thinks by developing the grasses, rivers can run the way they used to, filling from the bottom up. Overland flow in storms can be clean and clear, not the mud that washes away now. He recalls springs with running water on hillsides from his youth. They all disappeared long ago. He believes drought is the natural state of the country, briefly interspersed with floods. We have to manage with that in mind. He has never seen a farmer who wastes water. "Give a farmer a megalitre of water and he will produce food. Give the city a megalitre of water and they will produce sewage." Not all of what Greg says sits well with many people but the positive impact of his farming methods are clearly to be admired.

Back on the road it was time to piece together some of what we had learned about the river and its people. Most people who live on the river, who see it every day, are in despair. They are the few who 'feel' the river, who know its moods. Others rely on what they have been told, or they extrapolate an encounter at one place and one time to what they think the river is. People who use the river from time to time want to know it is healthy. The Surat fishermen understand the situation but are too polite to say why the river is so bad. As they say, "We have to live here." But in their hearts they know. It is the over extraction of water and the chemical runoff. Their reason for optimism is that much less of the poison is now reaching the river.

Nearing the top of a hill there was a white horse grazing beside the road. Behind it were about a dozen cattle. As with all the horses I had encountered, the kayak made it jittery. The horse stared for a minute and then trotted away, only to stop after a hundred metres, stare again, and run away again. The cattle turned and ambled back up the road the way they had come. From the top of the hill cattle were spread along the road for at least three kilometres. The horse continued its cycle of staring and trotting away. The cattle started to trot to keep away from the kayak. There were a few hundred of them, all turned around and being herded in the wrong direction by the time I reached the drovers' camp. It was two hundred metres off the road so when the horse headed that way I was able to get past it.

About a kilometre further on Steve, the drover, came over on his quad bike. "Reckoned I'd seen everything drovin'," he said, "but never seen anythin' like this." He had counted the cattle that morning: 1460 head. The mob was to rest here, having just come off a north-south stock route. My appearance had sent them back the way they'd come. We agreed he would ride ahead and bring back the cattle that I had sent off in the wrong direction towards me. With my trusty kayak I had a thousand of them heading the wrong way, west towards Surat, by the time he had turned the leaders around. Imagine 1000 cattle being herded in one direction and 460 coming head on from the other direction. When the two opposing forces were about to collide I pulled off the road as far as I could. Luckily Steve turned my mob around and all of the cattle continued past me in the right direction.

Chapter Eight

Ken and Barb caught up at this stage and got some good shots of Steve the drover and heard stories about droving life, how all his woes were caused by Peter Beattie, the then Premier of Queensland, and what we needed to do to 'fix those bastards'.

There are many tales about the damage caused by hard hooves. I decided to examine the effect of these cattle here. To compact soil for construction of roads or buildings there is a piece of equipment called a sheep's foot roller. This concentrates the load onto the tines, thus getting greater compaction. Given that as a model, one would expect the soil to be packed very hard after nearly six thousand hooves. The cattle had actually broken the ground up. It was moist from the recent rains and soft to kick into. I contemplated that the sheep's foot roller is only effective as a compactor in the right soils and at the right moisture content. Too much moisture and it squishes down and displaces the mud, too little and it just pushes the dry dirt aside. Clearly, many variables determine the impact of any particular activity. It was another lesson for a kayaker who knows zip about the country.

In Surat, Ken had made the camp site nice and cozy. My spot was in the back of the ute with a tarp thrown over it, and Ken and Barb slept in the big tent; the cooking and eating was under the awning off the side of the trailer. With a cold wind blowing and some rain and sleet, Ken had taken another couple of tarps out and provided some walls for shelter.

He wanted another satellite phone so that he could have one in the ute and I'd keep mine in the kayak. He had noticed that sometimes, even with the high gain aerial, the Next G phone had no signal. What would it be like down near Louth or Tilpa? Farnie had been good enough to support us with one satellite phone plus the UHF handheld unit which Ken and I were using for kayak to ute communication. Its range was five kilometres in this country. I rang Carol to get her opinion. We agreed that it was unfair to take advantage of Farnie and ask for another phone.

Dinner was at the pub that night. Again there were stories that this cold weather should have started seven weeks ago on Anzac Day. The fish restocking club had a notice board with lots of photos and stories. Each year they take thirty to fifty thousand native fingerlings and place them into the creeks and rivers around the shire. This group engenders a great community feeling. In 1983 I had studied local government engineering. They didn't award me a local government engineering certificate though, as I only had three months' roads experience with the other twelve years being water and sewerage. It would be a long time before the local government engineers woke up to the fact that there was more to local government than roads. Perhaps my letter to the reviewer telling him he was a dinosaur who needed to understand the significance of water didn't help my case – but TEFE.

The course taught us that there were too many local authorities in Queensland. These needed to be amalgamated into regional councils to gain efficiency. All students accepted this conventional wisdom. But in Surat, a fish stocking club had different ideas. The local council was about to be swallowed up in the State Government's forced amalgamations to bring about efficiencies all over Queensland. One of the members gave his opinion. "You see that road the council is working on near the airport? You will notice that everyone is working hard and that they are trying to achieve the best possible job. This is because they all live here. If they slack around and do a poor job they will cop it next time they come in here for a beer. So they don't slack around. They have pride in their town and their work. What do you think is going to happen when we have a road gang and engineer from Roma come and do this sort of work? Will they have the same community pride?"

With only 25 kilometres left to walk into Surat it should have been an easy day. A strong headwind and long hills of up to five kilometres made sure there was still plenty of challenge though. The wind was bitterly cold. Even at a spacing of twenty to fifty metres, trees made a huge difference. The wind is much less among the trees than on the stark, exposed plains. Regardless, 25 kilometres is a short day so we had from 2.00pm onwards to visit the museum, paddle on the weir to a reception of grey nomads, and investigate downstream of the weir.

The Surat museum indicates that from 1860 to 1890 Europeans made huge differences to the landscape. Life was hard and very expensive but it must have been profitable for some. Was this the start of massive land and water degradation? Local mayor Donna Stewart says Professor Peter Cullen, now deceased but a lifetime fighter for river health, told her the shire was the only one with pristine catchments. To me a pristine catchment is one that has its original purity. Donna is rightly proud of her area and her community but how can it be pristine, unless maybe the museum got it wrong?

Water flow over the weir looked good. Townsfolk had gathered excitedly a couple of weeks ago to watch water flow over it for the first time in a long time. For this kayak paddler, though, the Balonne River was disappointing. Just 500 metres downstream of the weir there was not enough water. The next water was supposedly 60 kilometres away by road. It was the tailwaters of Beardmore Dam at the Warroo bridge.

We had used the screen and projection gear back in Dalby but hadn't got it out for schools. Surat would be the first school where we tried it out. The morning was bitterly cold but the kids seemed to take it in their stride. We showed off the kayak and then went upstairs. Everyone crowded into a room where we ran the PowerPoint slides which showed the map of the route and touched on global warming for the older students. Surat is a combined school with all age groups.

The wombat meme finished the presentation. A meme is the name give to cartoons that express an idea. The name comes from the concept that there are individual units, like genes, that make up culture. Cartoon memes have become a popular form of communication on the internet. This cartoon was produced by an American called Jason Ables. The wombat is an Australian animal but Jason's looks more like a rat; and his voice is American. Overall the effect works very well and has an international flavour. It starts with: "Yo, listen up, this here's your home. The only one you've got." (picture of the world comes up). It goes on to talk about how everything is connected and we have to live together on this world. "One world. Not two worlds. Not three worlds. Just one."

The Surat kids loved it. The wombat is a powerful little guy that I used whenever possible. I had been introduced to him by Ruth Rosenhek.

The school presentation meant that walking didn't start until after 10.00am. Ken's web notes for the day were:

> We dropped into Surat State School and gave a presentation to the whole school before Steve set out on the road to St George. It was freezing cold for the entire day. I'm sure glad I didn't have to walk.
>
> I spent two hours downloading film …… in the heated public library, while Barb stood in the freezing cold cooking a delicious stew for us all. I picked Steve up after 27.9 kilometres of walking and we returned in time for dinner and a quick interview with Don Thon at the Royal Hotel.
>
> Surat has plenty to be proud of. They are community minded, have made a real difference to their local environment, have some very interesting and well put together displays of their local history and culture, have a very well run school and are all very friendly. Thanks Surat … we'll be back!!!

Ken is right about Surat: it has a great atmosphere. Maybe writing the bit about walking was a way of apologizing for yesterday's comment. When you're near the limit of your physical capabilities, issues like that are too hard. It has to be simple. Perhaps it had been a mistake to get Ken to back up the tapes on the computer. Maybe the time and stress of that was causing his problems.

He was certainly on edge at times. That night I said to him at a time of jest but also in a way that let him know I meant it "I will never hit you unprovoked, but if you hit me, make sure it is a bloody good one. Turning the other cheek is not my style."

An old bloke in the pub reckoned the cold weather we were experiencing is what the winters should be like all the time. His opinion was, "Now the cold is an aberration, so something is not quite right."

An email came through from Vikki Uhlmann. Vikki had looked after the Ipswich Forum, provided us with some climate change handouts, and was organizing all of the council contacts along the way. This was complementary to Jenny Cobbin's school's role. So far everything was working perfectly. Vikki had accepted a voluntary role with Engineers Without Borders so she would be heading off to Jakarta.

St George – Bourke

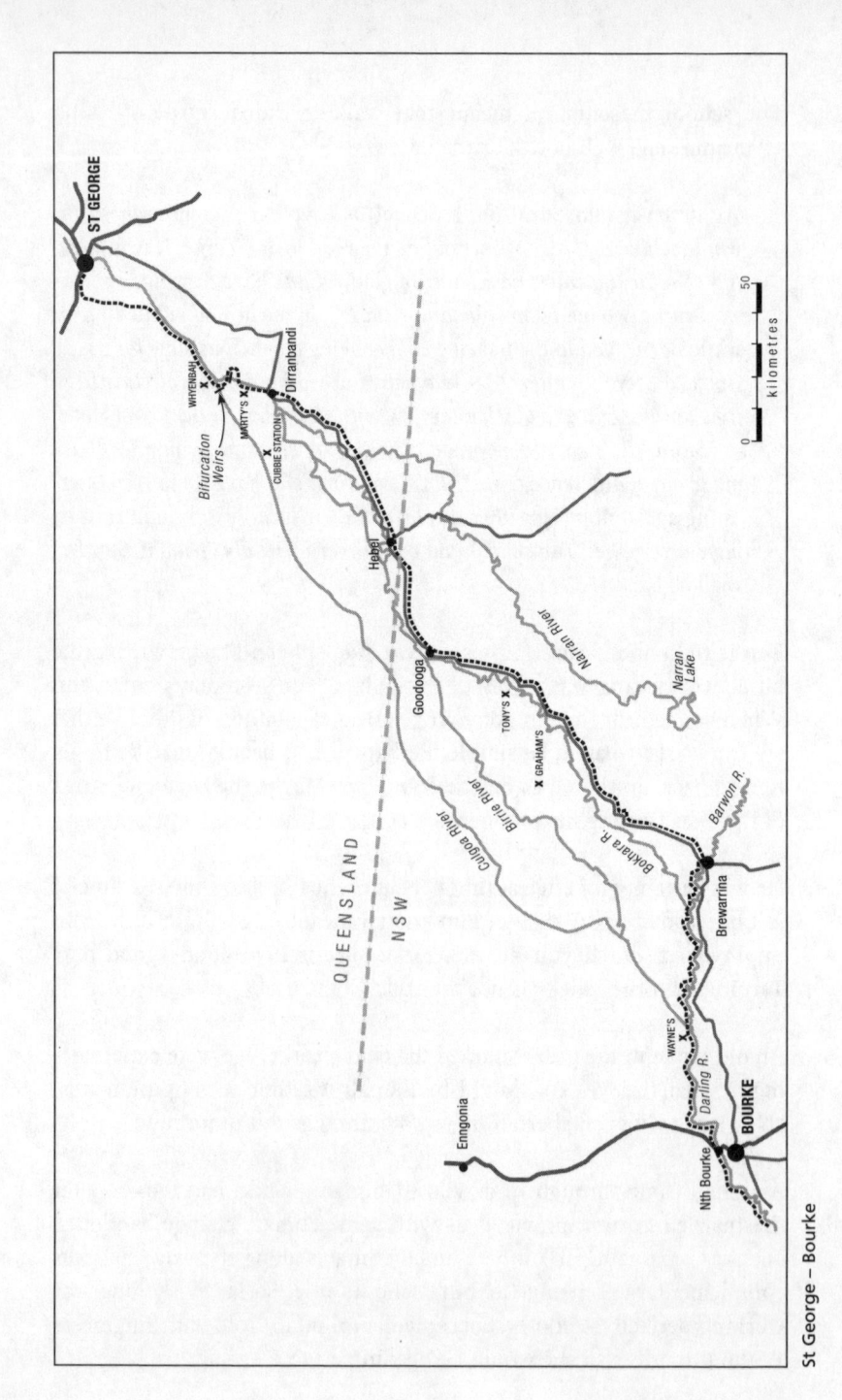

Chapter Nine

St George

THE TRIP DOWN TO WARROO WAS UNEVENTFUL, cold, bleak and punctuated only by a lukewarm Chiko roll delivered by Ken and Barb on their way to check out our accommodation, which was to be the old shearers' quarters. Our hopes for water backed up from Beardmore dam were dashed by the dry, sandy river bed.

Although the shearers' quarters were very basic there were real beds to put our sleeping bags on, with a real floor to walk on. When I got up in the middle of the night I fell back down on the bed. My feet wouldn't support me until I put the weight onto them very slowly. Five toenails were black, there was a huge blister on my little toe and two other toenails had fallen off. Only two toes looked more or less normal. This was not the problem that made me fall over at night, though. There was no strength between the bones along my feet to provide lateral stability. The reason for this was a total mystery, but that's why I fell sideways.

By the morning my feet were good enough for another day. The stay here had been organised by Jenny Cobbin and she had posted the letters from the Rangeville School students to us care of Warroo. We all loved reading them before going off to the Wycombe School. It was the last day of term. There were only four kids enrolled but a few more from Brisbane came along for a look. Some mums were there too.

Before leaving we chatted, as one does in the bush, talking about how they see the land. There seems to be lots of information around for informed people. They don't rely on help from a DPI (Department of

Chapter Nine

Primary Industries) official like they used to. There are videos, the internet, printed literature and seminars, all with a host of information about how to run properties. Despite this some are very successful and some struggle. Some of their stories reminded me of one Bil told.

There are two brothers around the Condamine area. One spends hundreds of thousands of dollars shaping his property. He has everything surveyed, drawn up on a plan and shaped with laser control to direct water where he wants. The other thinks this is all rubbish. Every time it rains he walks around his paddocks and looks at what the water is doing. After the rain, if he has found something he does not want, he goes out with the dozer and reshapes, ready for the next rain. Both are right. Both are effectively managing their land in a way that retains the soil.

A farmer pulled up in his battered ute. "Tell 'em not to pick on Cubbie," he said. "There is no water there. They have none at all. It's those bastards at Dalby. Gotta wait for Dogwood Creek or the Maranoa to run here. Can't rely on the Condamine anymore. Nuthin gets past Chinchilla." I was not surprised that Cubbie Station had come up. The reason I had come via Dalby and the Condamine was to see the most famous cotton farm in the world – Cubbie Station. No matter where it was mentioned, Cubbie Station polarised opinions.

My Achilles tendon was not too bad so maybe acute attacks were in the past. My infected finger was back to near normal size. Why was I tired? The strong, cold head wind didn't help. The long hills slowed me down because uphill was slow and to speed up too much downhill was to risk Achilles damage. Maybe a rest day was called for. Two days had been planned in St George but when Ken and Barb took time off to go to Sydney I could stop for a long while. Jenny Cobbin had suggested that I use that week as an opportunity to write. A week off didn't feel right though, my heart wanted to press on. A stubborn, relentless pressure to go forwards had taken over.

I thought about where I was. Finally the city shackles had gone. This was a different world. It was a world where the city was just a far away place; where the cycles of the moon are with you; where the stars shine brightly. Walking along lengthy straights, to use Ken's expression, you were 'just in the moment'.

In a phone conversation I had said to Bryce, "You know the stars and the Milky Way we used to see as kids?"

"Sure," he replied.
"Well, they are still there."

Today I long for those night skies with the Milky Way smeared across them. We forget too many things in a city. We forget what is important.

Sometimes I thought about where I was going. The adventure through the remote country past Bourke would be exciting. This is the outback of Australian folklore as well as the country hardest hit by the drought and water crisis. The Murray would be something different. Huge irrigation areas and a body of water many times that of the Darling, but no less damaged by our mismanagement of water. Because it is the water supply for Adelaide and beyond, it has a critical role in the national water management plan. Despite all this, overwhelmingly I felt the call of the Southern Ocean. I wanted to see it, to paddle on it, to feel it.

When Ken picked me up and took me to the caravan park at St George he had already set the camp up. This included my tent. So far I had slept in the back of the ute but here I was to sleep in the tent. This was great – my own space that I could keep for a few days.

Ken had called Farnie to ask for another satellite phone. Farnie had readily agreed, saying he had expected we would want it. I was uncomfortable about the request and angry that Ken had gone ahead despite my decision not to. As things stood, though, I only had time to deal with the practicalities. Amanda and her partner Matt were coming out for the weekend so I asked her to stop at Farnie's to pick it up.

Robert Buchan, the mayor of St George, invited us to the local Chinese restaurant for dinner. We met him and his wife Desley there. Don Alcock from Keytext had come out from Brisbane to do an interview

Chapter Nine

and soon two of my mates, James and Rory, arrived so it was quite a gathering. Robert picked up the bill for everyone, which was very generous of him. The following morning he turned up at the camp site with his own chair and a loaf of bread that he had just baked. He's a real character with all sorts of quirks such as an old Ferguson tractor collection, an old Bedford truck collection, a selection of number plates that he thought he would on sell at some stage and make squillions, and an incredible habit of reciting a person's phone number when he says their name. That one certainly saves on phone books.

I liked Robert. He's a climate sceptic despite the influence of Desley, who tried to get him to read *The Weathermakers,* which had convinced her about climate change. We disagree on a number of things but his hospitality and his love of St George are indisputable.

Desley is a doctor and wants to buy a scooter to get around as her contribution to reducing greenhouse gases but Robert will have none of that. Desley's professional advice to me was that she had seen many snake bites and if I got bitten to sit down and get someone to come to me. It was winter and I reckoned snake bite was unlikely but valued her advice. Maybe one day Robert will listen to her on climate change.

James and Rory had decided to come out that weekend as it was about the limit of where they could travel to from Brisbane. Rory and I had worked together with Greg at Aquatec in 1994. In 1999 when we were establishing Watergates at our house, Carol would provide lunch on our veranda as a way of getting to know our customers. One day, Rory and James were taking a long lunch with us – long enough that Greg would most certainly not have approved – and Carol told them about our exploits in Libya in 1978. A few months later both of them were off overseas to see the world. Greg lost two very good engineers, partly as a result of that lunch and the powerful urge to experience the world. Rory has only recently returned home to settle down so he had been away for a long time.

After we tucked into the mayor's bread, he excused himself and we arranged to meet at the park on the river at 2.00pm where he would organise some locals to welcome me coming off the water.

There was a call from a woman at ABC Radio in Brisbane wanting to know if they could have some photos to put on their web site to complement the story they had done. "Sure," I said, "Just take what you want off our web site."

"You tell them they're copyright Ken McLam," Ken said. I thought he was joking. "No, I'm serious. They have to put my name on anything they use."
"Sorry," I said into the phone, "did you get that?"
She was surprised but said "We can put an acknowledgement on what we use." "Thanks," I replied meekly.

I put the phone down. Ken said "I know about copyright. As a school teacher I work with it all the time. They have to put my name and copyright on any photos they use."

This was getting uncomfortable so I let him know. "Well, if you feel that way we had better talk to Geoff and see what he thinks about the web site."

What transpired is that each day Geoff put "Images by Ken McLam – Refer to copyright notice", with a link to a copyright notice with the legal information on what could be done with the images. I was determined to make Ken happy by offering this copyright protection but I was very offended. This was not the spirit of the trip. We were here to do good. After the trip it would be good to make some money so that we could continue to do this sort of work but the bottom line was that we should be friendly and helpful. A copyright notice was entirely the wrong image for the web site.

I emailed our solicitor in Ipswich asking whether this was necessary and what was the law? The response was that this sort of thing had some natural copyright protection under common law but a copyright symbol and a copyright notice strengthened the case. This information I copied to Ken. But I had another issue. Did Ken own the images? Sure he took them on his camera, but wasn't there more to it than that? He

Chapter Nine

and Barb had been picked up at their house, been transported the whole way in a K4e vehicle, K4e had paid all of their expenses including all drinks and even family phone calls. K4e was even going to provide a vehicle so they could go to a wedding in Sydney. Didn't K4e have some claim to ownership, some rights too? It took six weeks to get that answer and even then it was only verbal via Carol. The conclusion was that, yes all of that was true, but anything like that is best sorted out without resorting to the courts.

Rory and James arrived so all of this distraction went to one side. Ken dropped us out on the road for the walk into town. We all wore safety vests so the team was very visible. That morning was our first breakdown. We got a flat tyre and when we tried to get the wheel off, discovered that the bearing had failed. Ken was looking around St George for a new windscreen because a road train had thrown up a rock that had made a large crack. He brought out a spare wheel and went back to his windscreen duties. Another couple of kilometres and Ken had to bring another wheel out. The front wheel had collapsed. That was three failures in one day. To date there had been none. "Well guys," Ken joked to Rory and James, "A flat tyre, a broken wheel, a broken wheel bearing and a cracked windscreen, all since you arrived!"

It was Saturday. Ken organised to have a windscreen delivered so it could be fitted on Tuesday. He then arranged for access to the water at Kapunda Fishing Park. This was at the start of paddleable water collected by the town weir. After a morning impressing Rory and James with our technical prowess fixing wheels, there would be an easy 11 kilometres paddle to meet Robert and his welcoming committee at St George.

Amanda and Matt arrived just in time to say hello before the paddle. Molly, their dog, had a good run along the bank. Rory and James jogged back into town. After I arrived at the park in town the local journalist had me posing for photos and then we came up to the group of people waiting with the mayor. Don Alcock was there and about to get plenty of information for his Keytext story. Rory and James stayed for a while before heading for Brisbane, Ken set up to get some footage, Amanda looked on and Matt made sure Molly didn't make a nuisance of herself.

The group offered a conventional warm welcome. Then the real reason for the gathering unfolded. It was a set up. Their arguments surprised me. To remind me not to take a backward step I sat on my kayak.

Here are some comments from the group:

> "You have to live here for years and years to make a sound decision but the first impression you see you will go back home down there and report on. And I don't think you have enough expertise or general knowledge of the area to make a statement or comment on TV."

> "We are in melons too and I know many blokes where you are talking about. They haven't had a crop for years. You can't criticise someone for taking what is rightfully his. Any man that had half a brain would get as much water as he could to fulfil his requirements."

This last was in relation to my comments on what I had seen at Charley's Creek near Chinchilla.

> "We are a very important part of this nation. Not just Queensland."
>
> "We need government off our backs. Look around, talk to the businesses, the fish and chip shop, and see how this place works."
>
> "Look at the percentage (of water) taken out at St George and compare that with the rest of the basin."
>
> "This area is managed right. We are not going to alter the Queensland part of this river. They have to look at it from there (the border) down."
>
> "South Australia doesn't want our water anyway. It is too salty for them."
>
> "You tell them in New South Wales that we have no water here. We haven't taken it. They have to look after what they do a lot better."

They were very defensive. They didn't want a greenie coming down the river and making a judgement on them. Tim Flannery and John Doyle had upset them when they came through filming *Two Men in a Tinnie*.

Chapter Nine

Richard Carlton (the journalist), had upset them, especially by confirming the idea that a story had to be sensational. They thought too many people wanted to sensationalise the worst drought in history.

Perhaps they had been poorly treated. Perhaps they were managing the water resources in a balanced way that suits the whole of the basin. That does not justify stubborn self righteousness though. To refuse to accept that a basin must be looked at as a whole is ignorant. To believe that as long as they are OK all is well, is selfish and narrow minded. If they are doing the right thing then a whole of basin review will confirm that.

I was not surprised at the end of the trip when someone in Adelaide told me, "I was at the mayors' conference and a certain mayor had it in for you." "Wouldn't be St George would it?" I asked. "Spot on," was the reply.

The best quote from the meeting was "I've been a farmer all me life." "Not yet you haven't," his mate quipped.

Walking back to the caravan park we noticed a rugby match. It was golden oldies. One team was the Condamine Codgers, who had made the racket we had heard in Condamine one night. The other team was the St George Frailnecks. St George had many more reserves and were posting more points. Just when we arrived they knocked the ball forward near their try line. This means play stops and there is a scrum. The referee seemed not to notice while the young bucks (under 50) from each team scampered down to the other end of the paddock. The older blokes just stopped. In the end the ref blew his whistle and called "knock on", so the young blokes had to run all the way back. Obviously the older chaps looked like they had needed a rest.

There are some strange rules in golden oldies. A short line out in real rugby means a reduced number of men in the line. A short line out was called at one stage and they all knelt down – all except the 65 year old hooker who was really short and thus carried dispensation. Even with that advantage though, he didn't get the ball when it was thrown in. St George scored and the scorer kicked the conversion goal. He was immediately banished from the field and forced to drink four glasses of port as penance for such unsporting behaviour.

St George has some very good port producers. One is so good that the label proudly says: 'Fucking Good Port.' Well you don't want to leave a potential customer in any doubt do you?

The game was played in great spirit. Although the next game was a tough, first-grade game, it didn't have anywhere near the atmosphere, support or camaraderie that this group of geriatrics enjoyed. Luckily, golden oldies is not played too regularly as no doubt the recuperating time from over-exuberance is quite lengthy.

Sunday, the next day, was a day off. We worked on repairs until lunch time when The 7.30 Report from ABC TV arrived to do some filming on the road and out on the river at Kapunda Park. Amanda had brought fibreglass and resin to fix the paddle but there were other bits of housekeeping that were necessary. At breakfast Ken hit me, calling me an 'ungrateful prick' because I didn't thank him for arranging the satellite phone that Amanda had brought from Farnie. It was just an angry punch to the shoulder, but totally unexpected. In any other situation I'd have hit back, or laughed in his face, but it was a long way to Adelaide and I was on my best possible behaviour. It's a bit hard to thank someone for doing something you did not want done!

I mulled over the exchange. I really felt that Ken was always happy to take, and had abused Farnie's generosity. The phone potentially had a role but was not likely to be used and satellite phones are very expensive. I wondered if I should have let Ken know that I thought he had done the wrong thing? I didn't feel angry. I felt frustrated that we were teamed up together and I had no idea how to deal with Ken and Barb. My efforts to boost their importance when talking to schools had resulted in being told on a number of occasions that 'You'd be fucked without us.' Should I have raised these issues? Was I abrogating responsibility by not confronting them? I came to the conclusions that we were so far apart in our thinking that to raise any issues risked a blow up. I decided it was better to keep trying to jolly them along.

Amanda and Matt left before lunch. They took home a couple of bottles of the self praised good port. Amanda and I make a great team. Sometimes she says "I hope I wasn't too brutal with you." But I have to know what I'm doing wrong and sugar coating never improves the message. Both Amanda and her older sister, Heidi, have spent a lot of time on the front seat of the double wave ski. They know when Dad says "Uh Oh," that something unplanned and perhaps painful is about to happen. Paddling out through the waves, the ski launches vertically through the crests. If you don't quite get to the wave in time the ski crashes through the lip, slamming the front person on the chest. As I explained to the girls, "This is good because it doesn't hit the old guy on the back seat." We've had a lot of fun over the years and the girls learned about reaction time. When Dad says "Jump!" you do it now. Not in half a second. That might be too late.

On Monday morning the 7.30 Report folk came back to the caravan park and interviewed both Ken and me. When the story finally went to air it was an excellent chronicle about the trip and our goals.

Travelling back through the area in February 2008 it was wonderful to see what a La Niña can do. All the rivers we saw were full and some had even broken their banks briefly. Word was that the Narran Lakes were getting a drink and the ibis would be back.

There are lots of issues to be understood and many of them involve history. Terry Loos had arranged for us to see both Sunwater and the Department of Natural Resources and Water (NRW). Both departments (of the Queensland Government) were very helpful and friendly. Undoubtedly irrigation contributes enormously to the area's wealth. It also contributes to animosity.

Beardmore Dam is controlled by Sunwater. The employee we spoke to said things could have been done a lot better in the past but now they have to work with the situation that is in place. Some farmers are irrigators, some are pastoralists, others are both. The town of St George requires water. Everyone is waiting for the new water management plan to be completed. Jockeying for positions to get maximum water was straining relations everywhere.

The irrigators own the water in Beardmore Dam. When water flows in it is allocated in what they refer to as 'the bucket' belonging to each irrigator. They can then have their bit released when it is wanted. This is sometimes not an easy decision. If the river is dry when an irrigator asks for an allocation, none of it might reach him. This ownership of water sounds irrational until you realize that these guys have to pay for the infrastructure. Even when they get no water they have to pay their share for the dam and the delivery mechanism. Making any judgment on what is right or wrong is fraught with danger.

NRW advises Sunwater on what water to release and when. They even have obligations to New South Wales which include releases to the Narran Lakes. Almost all releases are limited to 730 megalitres per day. This is the amount that can flow down the river without 'wasting' it by spilling over the banks. The NRW guys understand the animosity water causes. It seems their work is one of the most unpopular jobs possible. It is easy to upset everyone and hard to please anyone.

The river system after St George is quite complex. The map on page 108 shows this. Details are not critical but it is important to know that it is like a delta. The Balonne River splits into the Balonne Minor and the Culgoa at Whyenbah. In turn the Balonne Minor splits into the Narran, the Bokhara, and the Birrie. The Narran flows to the Narran Lakes, the Birrie runs back into the Culgoa and the Bokhara flows down to the Barwon which heads west and becomes the Darling where the Culgoa comes in. At the top split from the Balonne to the Balonne Minor and Culgoa there is a weir on each river. These are steel sheet piles driven into the bottom and are about two metres high.

CHAPTER NINE

On the Culgoa weir there is an adjustable gate that slides sideways to allocate the balance of flows between the two rivers. This is a Watergates product, installed in 2006 when the boss was in Alice Springs Hospital after crashing his motorbike on the Plenty Highway.

At Watergates a few years earlier, there were some big jobs around for gates on cotton farms. I had a problem supplying to such locations and thought that we should only supply gates where we knew they were for environmentally sound uses. This could extend to fish transport over a dam even if we disagreed with that dam, but it couldn't be extended to irrigation works where we had no idea if it was a sustainable use. When it was explained to the staff that this could mean reduced job opportunities and make life tough for us at times, they still had no hesitation in doing the right thing. I was very proud of our team.

We took the opportunity of the rest day to interview Robert at his workshop. Here we looked at his collection of old trucks and tractors. Despite our difference and knowing that he had set me up as raw meat for his mates on the river bank, I liked him a lot.

Like all mayors he has plenty to say and is immensely proud of his shire, where he has lived since 1948. Perhaps his key message is that too many city people think of country people as farmers but there are whole communities and businesses based around the farming enterprises. He believes all farmers are conservationists at heart. They provide food and fibre for the nation and city people need to be a little more grateful, or at least understanding.

Back at the caravan park we found Bill Gorman, president of the Murray Darling Basin Association. Bill was on holidays from Victoria but was talking with Queensland councils. He takes a more holistic view of the basin but is very aware of the role of local populations. While agreeing strongly with the need for whole of basin control, he says "It is all very well for the powers that be to make decisions but they must listen to the people. It is all about the local people who make their living out of the water. They have to listen to them if they want it to succeed." Too true, but if the local people will not entertain any discussion about a whole of basin strategy, then any controlling body needs to have the teeth to enforce such a program.

On Tuesday it was back on the road. The bridge out of St George is also the town weir. It is narrow and over a hundred metres long so it was necessary to get over it as quickly as possible. This is easier said than done when you're a gates man and there are interesting control gates on the weir. At least one photograph was required.

Barb was up ahead in the ute, ready to wave to cars telling them to slow down. Cars are not the problem though. It is the road trains. She waited a few hundred metres on, where the road turns off to Whyenbah, the direction we were headed. When I was a hundred metres away she turned around and headed back to St George giving me a friendly wave as she went past with the windows up. There was no chance to say thank you, no verbal exchange. The same old feeling of disappointment was there. It bugged me that despite trying to rationalise those emotions, I was no further ahead in getting the team dynamic to work than at the bottom of the Toowoomba Range.

The bitumen stopped after a few kilometres so it was back to gravel again. Our neighbour in the caravan park drove towards me. He had been down the bitumen to Dirranbandi and back up the dirt. He said he had seen water and would show us that night. He departed and another two hundred metres down the road there was a bang, like a rifle shot. The tyre had exploded and green fluid was all over the gravel. After I called Ken on the phone he arrived with a spare wheel and the new bearings. The bearings didn't fit so that was it for the day.

That night, after viewing the neighbour's photographs of water, it rained about five millimetres. Rain on the tent was a great sound. The morning temperature was about five degrees which was decidedly warmer than we were used to. We went back out to the road which had some puddles on it but seemed to be drying out. The moisture made the kayak harder to pull and while it was warmer to get out of the sleeping bag, the temperature stayed at five degrees for

CHAPTER NINE

a long time. With a thirty knot wind in my face I was careful to keep my hands tucked up in my jacket sleeves.

There are a lot of very big ring tanks in the area. There are also huge channels to move the water around. Running off these are smaller channels filled via control gates on large pipes through the walls of the channels. The scale is staggering. None of this had any water in it. The large ring tanks still have some trees in them so to climb up to the top of the embankment presents what looks like a paddock with a monstrous dirt wall around it. Perhaps this demonstrates that water in these tanks is only ever transient. Perhaps it means that these are new and yet to hold water for any length of time. With an evaporation rate of 2.5 metres per year it is unlikely that any farmer would want to store water for very long – maybe two crops at most.

At noon I called Ken on the satellite phone to give him my co-ordinates. This was just a precaution we had agreed to in case anything untoward happened.

The spot described by our neighbour soon came up. I left the road, walked 500 metres over to the river, slithered down the bank and set off into the brown caps. They would have been white caps had the water not been such a dense, dirty brown. Although a 30 knot headwind is not usually a lot of fun, today it was enjoyable because it beat walking. Three kilometres further on there was a sand bar but the farmer had graded a small channel through it, allowing the kayak just enough passage. Another three kilometres downstream there was another one but this time there was no channel. After a kilometre of exploring down the river with no water in sight, I decided to cut my losses and get back on the road.

Slithering down an eight metre muddy bank is easy. It is rather more difficult to slither up one. After much grunting, lifting, and slipping backwards the wheels could be lowered and I marched out to the road via two shallow gullies. We had talked about me camping along the road by myself but with rain threatening, I decided to postpone the idea. To carry my camping gear would have added another 15 kilograms to the kayak so, struggling up the bank, I was very pleased we had decided to delay my solo camping.

The road was getting busy. Three cars came past in only an hour and we had the obligatory chats about the trip and their farms. Ken picked me up at 5.00pm and we went back to the caravan park for a communal barbeque and sing along with our grey nomad mates. I had only covered 30.5 kilometres including the six kilometres paddling, but Ken's day was worse. He went to get the windscreen fitted. It had been due to be fitted the day before but had been sent to the wrong town. Now it was the wrong size. When he picked up the new wheel bearings they were the wrong size too.

There was another bit of rain that night and when Ken drove me out to the start point in the morning there was a grader on the road. Enough rain had fallen to collect puddles in the gravel so the top could be graded off, turned over, and was mixed to something that could compact to a good running surface. The loose gravel was about 50 millimetres deep so the ute moved around a bit. Ken commented that it was too dangerous to tow a trailer on that but I discounted what he said, believing that when he drove back he would realize it was a bit of a silly statement.

One of the back wheels on the kayak was down slightly but not much. We had no spares as they were at the bearing shop. There was not much we could do except maybe pump it up tightly and risk a blowout like the day before. Here is Ken's log:

> *Dropped Steve 30 kilometres down the dirt. As we unloaded the kayak I indicated that the rear tyre was looking a bit 'suss'. "No worries, she'll be right." Posselt sent me on my way. The grader had been busy mixing the large puddles of water and the soil into an interesting slurry so 4WD came in handy.*
>
> *Just as I hit the bitumen Steve rang on the satellite phone to say that the tyre was totally flat. Barb and I packed up camp and dropped in to pick up the new wheels with the new bearings and ... they weren't ready. They'd been sent to Brisbane instead of St George. After dropping*

> the trailer at the Dirranbandi Caravan Park I removed a wheel from one of the spare kayaks and we headed up the Whyenbah Road to find Steve.
>
> We found him huddled round a small fire in the lee of a bush sheltering from a very cold wind. After replacing the wheel Steve set off and only managed 10.44 kilometres for the day before we headed for Dirranbandi along the black soil road with rain clouds threatening.

After Ken had dropped me off a woman drove past whom I had talked to the day before. Her mother was at Whyenbah Station and she drove her little car down the road to see her each day. There were a few cars up and down the road during the day. Given the traffic I assumed Ken would turn up with the trailer about an hour after I called but he refused to bring the trailer down the road. It took almost four hours for him to arrive with the tyre.

During that time a real estate guy stopped. He had taken his old Holden down to Dirranbandi and was on his way back. He gave me the St George paper saying "You'll like that." Most of the paper was devoted to the golden oldies rugby match and there was a good article about our trip. The brother-in-law of the woman who visited her mother every day stopped and told me he had just qualified as an 'oldie' by a month. He was one of the tearaways who ran up and down while the truly old guys had a rest, and was featured as a bit of a star. Despite his youth he was feeling the worse for wear.

Sitting on my bum for four hours I worried about what we would do when the going got tough. I had no doubt that we would be in places that tested 4WD vehicles, and on top of that we would have to drag a loaded trailer. This was a lonely situation. There was no one that I could talk to. It had been established in earlier team discussions that my job was to keep up the morale of the crew. Ken would be way too sensitive to talk to about it. So I decided to postpone any action until after Bourke, when I thought we would need to confront the differences between Ken's approach and mine.

CHAPTER TEN

The Last of Queensland

DIRRANBANDI, ABOUT 90 KILOMETRES SOUTH of St George and at the start of the floodplain country, is the closest town to Cubbie Station. It is close to the border of New South Wales and Queensland and has featured regularly in the news because of water management. It was my intention to get a solid first-hand grasp on what was happening out here. The sensational headlines and passionate opinions over cotton farming do not really shed any light on Australia's water problems but emphasise the need to find solutions. With my experience I hoped to get to the truth.

In Dirranbandi we met a lovely couple, Marj and Sam McClelland, who were driving their tractor to Biloela for a tractor show. They had come from Mannum on the Murray River in South Australia. Eventually we would pass that way but I didn't give it much thought. The old Chamberlain tractor is very noisy so they wear ear plugs and communicate in sign language. It has a cabin and there is an extra seat for Marj and she seems very contented with her lot. A fold up camper trailer is towed behind. They made it to Biloela and back home again without a problem. This I know, because we were to meet again in a few months.

From our base camp at the caravan park we ventured around the corner to the hotel. Pubs are always great places to meet people and learn things. There was a young bloke called Marty there. He was cocky and self assured but very likeable. His place is not as big as Cubbie but he had just planted 6,000 acres of wheat, hoping there would be enough rain to make it viable. This was on just one of his many paddocks so obviously the place was not too small. We talked about the dirt roads with no gravel. Marty advised us that if you get 25 millimetres of rain you grind and slide your way through. "You can make it through, though?" I asked. "Sure," he said, "But you make a bit of a mess."

The camp site had a grassy section where Ken and Barb could pitch their tent. The trailer and ute were on stones but pegs could be hammered through them. With the possibility of rain, Ken had taken the three tarps to cover the awning on the trailer. There wasn't one for me to put over the back of the ute so I slept with my feet hanging out. Because it didn't rain the lack of roof was not an issue.

We had taken to checking the weather on the Elders web site. No doubt the information comes from the Bureau of Meteorology (BOM) but the Elders site gives a bit more local detail than the BOM site out in these regions. The site indicated that we had a 50 percent chance of less than one millimetre of rainfall. These statistics seem strange but the way I interpret them is that there would be showers around that were delivering less than one millimetre of rainfall as they passed and that the chance of one passing over us was 50 percent.

Ken and Barb drove me back out to the road but today would be different. There was much to do, people to talk to and important features to see. For starters it was only three kilometres down the road, around the end of a ring tank about a kilometre long, along a track and into the water backed up from the bifurcation weirs. Bifurcation is a Latin word for splitting into two. Where the river bifurcates and there is a weir on each branch, the weirs are known as bifurcation weirs. Most of the locals refer to them as the 'burification' weirs. This difference in pronunciation is also true of the Culgoa. Most locals call it the 'Culgo'. Maybe these local pronunciations are one way in which new words form and the language grows, but it is disconcerting.

After helping me unload the kayak, Ken and Barb waited at Whyenbah bridge. At the start of a day's paddling, Ken and I would lift the kayak down from the roof and load it with the drink bottle, food bag and

GPS, all rope-strapped to the deck. I would put on sunscreen, lip balm and make any special preparations we thought might be needed for the day. An important one was to check that the toilet paper was still dry. Another was to make sure the satellite phone was in its compartment. Ken didn't help push the kayak. Because the section from the road to the river was part of the journey from Brisbane to Adelaide, I had to do that myself.

Whyenbah was supposed to be less than an hour's paddle away but there were some logs and branches that slowed progress. Ken had a fire on the sand when I arrived and, despite the wait, they seemed to be in good spirits. They headed off to Whyenbah Station, down the western side, to get directions to the bifurcation weir on the Culgoa and I set off on plenty of water to do the same.

The wind out here is different from back on the coast. It is more variable and I found it difficult to predict. A windmill on the bank would spin madly, slow down, turn through more than ninety degrees and then spin madly again. Such a wind change does not matter much on a small river. Wind either funnels down the river or blows across the top. Across the top is good because it is calm against the windward shore. Wind from behind is great and wind 'on the nose' is hard work.

Ken called to tell me he had met the owners of the property and one of the guys was taking him to the weirs. He had already been past the junction and I was to take the left fork. It was good that these radio conversations were brief. With this one I was only blown backwards two hundred metres while I talked.

Arriving at the junction of two channels is like nothing you would expect. 'Am I here?' I wondered. There was a gap under a tree that had water in it. The tree was about a metre above the water and the gap between branches was slightly wider than that, but not wide enough to take the paddle sideways. On the other side of the tree, the water seemed to head off about two metres wide. Ducking under the branches I headed along what I hoped would be the Balonne Minor but it was a long way from being impressive. In a few minutes Ken and Barb and their new friend, Tom, came into view.

Chapter Ten

The weir isn't much, just steel rammed into the river bed with some concrete either side. If you stand on the concrete apron on the downstream side you can almost see over it. There was no water below the weir so it would be the end of paddling. Upstream of the weir is a structure that holds the stream gauging equipment. There is a track to it but I had also seen a track heading east. I wanted to know if I could get to the road from Dirranbandi by taking the kayak along the track. Tom didn't know where it went. His excuse was that he was not quite a local. Even though his family lives here at Whyenbah, he had moved to Toowoomba in 1987. Tom reasoned that all tracks eventually lead to a road somewhere. When pushed, he thought it might swing south a bit and then head along the property boundary to the road.

We decided to go back to the weir on the Culgoa. The others had walked past it to get there and by river it was less than a kilometre away. On the Culgoa side there was barely enough to paddle on. It was well worth the trip back, though, because here was one of our water gates. The purpose of this particular one is to control the ratio of flow between the Balonne Minor and the Culgoa. It is three metres wide and half a metre high. It slides sideways when a wheel is turned on the bank. Tom was unimpressed when I explained what a great engineer had designed the gate. Ken had already told him that I had done it. Tom seemed to think it was too hard to operate and should have been electric or hydraulic. He's a nice enough bloke but obviously cannot recognize talent when he sees it.

We learned, some days later, of a party trick of Tom's. When he gets pissed he likes to impress people by climbing up palm trees. People have many ways of showing when they've exceeded their sensible intake but this one is at least novel and entertaining – provided it does not end in tears.

My plan was to paddle back around to the other weir and strike out on the rough track hoping Tom's guess was right. I suggested that Ken and Barb drive to the farmhouse near the eastern bank and ask the farmer which track to take to get to the gauging station. If no one was home then it should be relatively easy to guess the way in. It would be regularly visited by government people.

Barb was unhappy about this. They wanted to wait out at the road for me. The difficulty with that was that I'd be setting out dragging a kayak on a track that may, or may not, go in the right direction. "You can call us on the radio and tell us where you meet the road," was Ken's advice.

I explained that it would be easier and safer for Ken to drive in and see where the track went. Ken complained that he might not be able to find the track. "Just go to the weir and follow it until you find me," I said. When Ken complained that he might not be able to find the weir I gave him the latitude and longitude. It was fortuitous that I had memorized the GPS co-ordinates of the weir. They had run out of excuses and agreed to find me. I was astounded at their attitude. Again, I had the strong sense that the support crew had a completely different understanding of their role than I did. I did not really feel supported.

The track was hard work. Recent rain had wet the ground to a depth of about 10 millimetres. This had then dried and cracked into a patchwork of squares about 50 millimetres across. It looked fine but there is a lot of energy used up in breaking the bond between these squares as the kayak makes its tracks. Dry sand is a lot worse, as is thick bulldust, but this stuff was hard enough that I could only pull about 500 metres without a rest.

At first the track went east, to where the road was. Then it headed south, hopefully to the boundary line that Tom talked about. But then it went north east, then north, then north west before finally turning back east. Luckily there was a sandy section that had been compacted and there was a brief respite from the relentless drag.

Near a low hill, the grass was more evident. A family of wild pigs was grazing on it. Stories abound on how fierce a wild boar can be. A yellow kayak is more than these animals can take though, so they were all still running into the distance when I lost sight of them.

After two hours the ute arrived on the track ahead. Ken and Barb had been to the farmhouse. The owner was very hospitable and gave them

Chapter Ten

a bag of lemonade fruit before directing them to a track intersecting mine halfway out to the road. Tracks are important in this country but not essential. With the backup of 4WD in case of problems you can weave your way though most of the bush. This is possible in a normal car but if something were to go wrong, like dropping into a soft patch, it would be a long walk to get help.

The ute was on the other side of a deep gully. Ken waved for me to stop while he set up the camera. It was a chance for a welcome rest. On the other side I continued on my way, stubbornly dragging the heavy weight through the cracked dirt. I was in some sort of zone. This single minded focus sometimes causes damage. There is a goal, which in my case was the road, and the person determined to prove their endurance just hammers on without thinking properly. It is where I'm at my worst – or in a fight at my most dangerous. 'Fuck the sore body, fuck the weight, fuck the dirt. Push on.'

The ground was opening up more. There were holes that the wheels sank into. A car wheel spanned them with hardly a bump but the kayak wheels are a lot smaller. Ken and Barb stopped ahead at a gate. Normally I would walk up to them and stop for a drink but I couldn't make it. I needed a rest before I got there. When I finally reached them, Ken said it was only two kilometres to the road. 'Thank God for that,' I thought.

Eventually I made it, turned right onto the road and Ken filmed me bouncing over a cattle grid. The grind across country had been hard but I was pleased to have completed it and looked forward to a beer at the pub with Marty. Ken and Barb told me they would wait five kilometres down the road. What should I do, tell them I'd had enough and call it a day? Only a sook would do that. I knuckled down and plodded on, stopping just 10 kilometres short of water backed up from the Dirranbandi weir. It was a very weary bloke who hobbled into the pub that night.

Rod and Shirley were late. They were coming out for the weekend and like Peter's, Shirley's food is superb. Rod has been an AWA mate since he challenged me to a 6.00am race at a water conference in Canberra in 1987. We've not had any races since but have spent a good deal of time together, including long meetings at certain footpath cafes, strategically analysing what is wrong with the world.

Rod drove down the same road that we had come down from St George, thinking he might catch us, but it was almost dark by the time he got there. It was an excellent feed, as expected. The next morning Shirley had bacon and eggs ready. Cooking was done in the camp kitchen where all of the farm workers cook when they are here.

The plan was for Rod and Shirley to walk the first 10 kilometres with me. Rod and I were lounging on my bed in the back of the ute because they were the only seats left. He couldn't believe how far it looked as we drove out to the start. "Don't worry mate," I advised knowledgably, "It always looks a lot further from the car."

"You know," said Rod, "the way you beat global warming is one step at a time, just like walking to Adelaide." He thought this was pretty good, adding "You can use that. I'll let you have it….for nothing."

My legs were tired and I had a blister on my little toe that was of some concern. It had swallowed the toenail and was growing inexorably.

We chatted about what I had seen and the reception in various towns. Rod and I have been mates for a lot of years and each of us knows how the other thinks. He's a very intelligent and capable engineer. What governments have paid a million dollars for a consultant to do, he can work out on the back of an envelope. Like most of us who think outside the box, though, the corporate world pays scant attention to his ideas, no matter how successful he has been. It is cold comfort to know that I was just one of many people who had been successful beyond the wildest expectation of the corporate management, only to be told that our methods were incorrect.

He and Shirley had no problem with the distance and went for a coffee at the end of it while I paddled down to the weir.

Chapter Ten

The river went past Marty's house. It was the best house that I had seen on the river. The banks are very low and the river overflows easily, hence the area is known as floodplain country. Even in the depths of the dreadful drought the water in the river was four metres deep. A sprinkler ensured there was plenty of green grass. The house was nestled in the trees. It is very comfortable and at the same time impressive. Marty had given me the farm UHF channel and came over from his fire to talk when I called him. Being Saturday he was setting up a barbeque a few minutes' drive away from the house. The family likes to get away like that.

Further down the river I heard an outboard motor behind me. Marty had taken the kids for a ride in the tinnie. It sputtered and stopped. A few minutes later he roared up beside, then took off again. Near the houses in town it sputtered to a stop and was still stopped when I caught up. He had a problem with the fuel connection but seemed to know how to fix it. We didn't see each other again until late afternoon when we drove out to his camp fire.

Back in Toowoomba I had heard that cotton shuts down above 36 degrees Celsius. Marty had checked his plants a couple of years ago after a 42 degree day and found three bolls had grown on one plant. He does not subscribe to the idea of temperature shutting down growth. This was just another example of the conflicting stories around. Truth is always hard to find and in cases like this, who knows? Maybe both ideas can be right.

The weir is downstream of the town but there is a water hole that extends to the bridge on the road to Bollon. Ken filmed me walking around the weir and then we pulled the kayak out at the bridge just three kilometres further on. That was the end of water for a very long way. Rod had his car so he and I didn't have to clamber into the back of the ute as we had done that morning.

With seats for everyone we could travel further so we headed out to Cubbie Station before going to Marty's. The Cubbie Station office was closed, because it was Saturday. We drove around hoping to find someone and marvelled at how pleasant and unimposing it is. It looked functional. I had an issue with sprinklers on lovely green lawn when, in Brisbane, what was left of lawns was brittle, grey and parched. There was no one to be seen so we drove further until we came to what we thought was the Culgoa feed to the farm. There is no doubt that the channel is huge. The control gates are each about two metres wide by three metres high and there are five of them. It all looks incongruous in a drought. All of the channels I had passed in the previous 80 kilometres look just as incongruous. When the rains come in Queensland, as they did in January 2008, these gates and channels are an amazing sight. The vast land is transformed to a myriad of channels connected to the river and giant storage areas. Together these storages dwarf Sydney Harbour many times over. When you see the country full of water it looks right, a clever domination by man over nature – but looks often deceive.

This is floodplain country, dry through much of its life with low average rainfall of only 350 millimetres per year, but submerged every time the waters come down from up north. Some years no waters come. Sometimes the water flows four times in just one year. Overland flow here means the rivers break their banks and flow out across the land. Back up on the Darling Downs, overland flow means rain falling on the land and flowing over that land to get to the rivers.

Cubbie people are sensitive to criticism. They cop a lot of it. They think that what they are doing is right and they see themselves as environmental guardians of the land. Their web site stresses the following key information:

- all water diversions are authorised by the Condamine-Balonne Water Resource Plan
- **River flows** - only a share of any flow is extracted, ensuring flows continue to the downstream river system
- **Floodplain flows** - the volume of floodplain flow diverted is water that would have been naturally consumed through seepage, evaporation and evapo-transpiration in the areas that are now levied

off from the floodplain. This extraction has no impact on water passing downstream
- Cubbie Group, on average, extracts only 0.28 of 1% of the Murray flow
- once the volume of water is extracted, it is then stored in deep storages to minimise evaporation
- the development is integrated into the floodplain to ensure the passage of flood waters to the downstream river system
- the development is a closed system ensuring no nutrients or top soil enters the rivers on the Murray Darling system

This looks impressive. Another way to look at the extraction percentage is to take into account Murray and Darling flow. The Darling has historically provided 14 percent of the total. Take 0.28 and multiply by 14 percent and Cubbie's extraction is 2 percent of the Darling. This is much larger than Cubbie's share of the basin's area.

The claim that their construction has no impact on water passing downstream is very strongly disputed by many people. To claim that the only water taken is what would naturally be consumed by the soil is a bit far fetched.

There is no doubt that Cubbie Station is a huge property but it is an amalgamation of farms and thus claims efficiencies of scale. It can certainly employ many people and provide significant income for local communities when it is operating. When there is no water nothing much happens. The Culgoa is its western boundary, the Balonne Minor and Birrie its eastern side. Both sides have a diversion dam about twice the size of the bifurcation weirs. Water is extracted only when the licence conditions allow. Between the giant storages there are floodways that allow floods to pass. The width of these pathways is less than half of the width of the control and storage areas.

Is it the monster that many people claim? Perhaps not, but the claims by other landowners about its overall effect on the floodplain make it clear that it does have a major negative impact.

The drive to Marty's was only 12 kilometres. There we had a beer and filmed Marty and his father-in-law, John, talking about their land and water. The property is 1,800 acres of irrigation, 8,500 acres dryland crops and 7,000 acres for cattle. The water licence is only for high flows coming past Whyenbah. Suction pipes are set at a high level so the pumps can only operate when the water reaches licence height. The pumps are sized at licence quantities so there is nothing to do except turn the pumps on when they will operate and advise Queensland Natural Resources and Water Department (NRW). Everything is recorded on a computer linked to the NRW office in St George. The farm had full crops in 2004, half a crop in 2005, in 2006 wheat reached 150 millimetres high and then had to be used for cattle fodder. In 2007 they planted 6,000 acres of wheat which was just showing through the soil.

I drove around the farm discovering that it is an enormous flat surface, interspersed with long earthen banks. Just one bank took Marty six weeks to build with a D9 bulldozer. This is a big piece of equipment. Everything on the farm is big. All the tractors are as high as a house. Fuel in 2006 cost over $600,000 for the year.

Much of the land has been cleared recently. After clearing and shaping there is a massive job to clean it up. "Stick pick'n'," is what Marty calls it. About twenty or thirty laborers, usually backpackers, are driven out to the paddock. They pick up all the sticks, logs and tree roots, placing them in heaps. At dusk Marty burns the piles. There can be up to five hundred fires dotted over the landscape. There is still one paddock of 2000 acres left before the lot is completed.

Marty faced tough times. Debts were huge and the wheat crop was a gamble. The farm is owned by Marty, his wife Karen, and his mother and father who have a property near Dalby. There were no handouts to buy it. Marty needed $250,000 cash, which he had made, and then leveraged up to his quarter share of the $5M price tag. Now it is worth $11M but there has been a lot of expenditure on equipment and land development.

Asked how he copes with drought and tough times Marty said he takes his father's advice. "Son, when times get tough, and they will, harden the fuck up."

Chapter Ten

The house is nestled beside the river. Green grass, trees and gardens make a picturesque setting. Away from the house, the might of man is starkly evident. Nature is gone and an efficient money making enterprise takes over.

Karen's father John was visiting. He farms in Stanthorpe, which is on the Great Dividing Range near the New South Wales border, straddling the watershed to the Clarence River on the coast and the Darling Basin to the west. Crops are capsicums, nectarines, peaches and apples. John is second generation Italian and proud of it. "Water is our life blood," he says. "We get more tonnage for less water than anywhere in Australia." On an apple tour of Victoria John saw practices that disgusted him. "Biggest water abuse I have seen in my whole life – open channel, sprinklers. Our computer controlled drip systems leave them for dead. Their water use is unacceptable. Everyone blames Cubbie but the irrigation on the Murray is ten times worse."

John may be a bit biased but he certainly tells it as he sees it. Much of the water he uses now comes from the sewage treatment plant. This recycled effluent must be used underground as spraying it is illegal. There are six farms in the area tapping into this valuable resource. John thinks the creeks in his area are in good shape. This is in stark contrast to Di Thorley's observations. The mayor of Toowoomba had told us that she remembered the creeks as permanent waterways, integrated into the environment and feeding springs, soaks and other oases. She firmly believes that farming practices since the 1950s have destroyed the water courses. John may not have seen what Di saw. Maybe he came along later so his different experience may influence his view. With such opposing views and controversy, is it really possible for a government authority to sort out a sustainable water management system for the whole basin?

Back at the house we had to protest to get away from the family's hospitality and we only made it after accepting a bag of apples with the different varieties explained. Rod, Shirley and Barb are wine drinkers. Ken and I drink beer. Marty had plenty of both.

The rig with the spare kayaks (page 20)

The welcoming committee at St George (page 115)

The Condamine near Dalby (page 84)

A log jam on the Condamine (page 98)

Bill's old bridge near Clifton (page 62)

Disappearing river – time to resume walking (page 73 or 92)

CRY ME A RIVER

Brisbane river – pushing the probe (page 31)

Gowrie Creek – into the unknown. Fast! (page 67)

CRY ME A RIVER

The bifurcation weir on the Balonne Minor (page 127)

Lonely camp – Dave loved it but how do we get wood for a fire? (page 154)

CRY ME A RIVER

Healthy and dying (inset) coolibahs at Tony's (page 153)

It took a long time to understand these large claypans (page 153, 195)

CRY ME A RIVER

A log jam on the Barwon (page 167)

Learning to ride the bank (page 166)

CRY ME A RIVER

Long stretches of the Darling are like a railway cutting (page 182)

Jonathan has the camp ready (page 192)

Finished for the day (page 192)

Bad algal bloom near Tilpa (page 194)

CRY ME A RIVER

Snags are essential for fish habitat (page 208)

The Darling River at Wilcannia (page 202)

CRY ME A RIVER

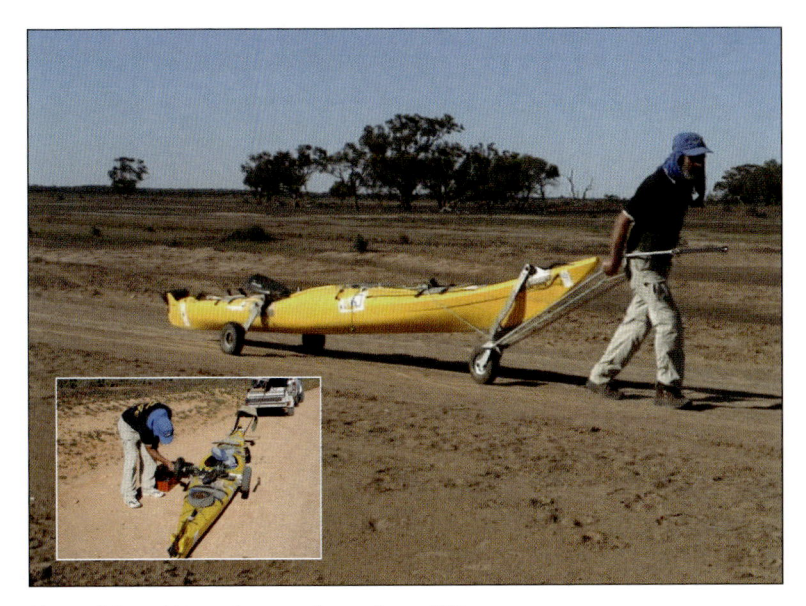

Dirt road. Inset: Yet another tyre change (page 121)

Darling serenity (page 208)

CRY ME A RIVER

End of the water at the Pooncarie weir (page 211)

At last, the Darling meets the Murray (page 217)

CRY ME A RIVER

Typical pumps on the Darling and its tributaries

Water sampling (page 220)

Going through a lock on the Murray (page 230)

CRY ME A RIVER

The pumps on the Murray are different again

CRY ME A RIVER

Bad weather brewing on the Murray but striking colours (page 245)

The Murray ends with a whimper at a man-made barrage (page 254)

That night in the camp kitchen, Shirley cooked up a storm again. We even had a choice of meals. After dinner, with the washing up done we sat chatting at the table. Explaining to Rod that this sort of life was very rewarding and a lot of fun I described my goal of going on the speaking circuit to fund such trips so we could continue to make a difference.

> Ken said, "Yeah, it's all right for you to make money. What about us, we do all the work."
> I replied that he could do the same. "You have a parallel trip with a different perspective. It would be interesting in itself, particularly the filming, so you could put together a presentation and do the same thing."
> "Well you'd be fucked without us! It's not fair that you get it all."

A lid came off something when he said that. He had said it a few times before. I had discussed it with Carol and a couple of the team and had agreed to jolly them along and play the adult but this was cancerous. It had to be dealt with one way or the other.

> "So what about when we get to the tough bits?" I said. "What happens if I'm out on a dirt road and it rains so that I can't pull the kayak? What if it is like it was back up towards St George with the rain but the road has no gravel on it?"
> "Then you sit it out. You wait for the rain to pass and the road to dry out before we come and get you."
> "We bought a $35,000 4WD ute for you to drive. I have a kayak that can't be pulled through stuff that an ordinary car has no trouble with. Any time I get stuck it should be nothing to pick me up and get out of there before more rain comes."
> "You are the fucking adventurer! Take the tent and all you need for a week and wait."

This was starting to piss me off. I had warned Ken twice that I don't turn the other cheek but had let him hit me and get away with it. This had to be resolved.

> "You are the support crew. That means you're there to support me. Will you do that?"

Chapter Ten

"I think Barb and I have had enough. You would be fucked without us and you need to know that."

"Well Ken, any decision you make is yours. I'll be disappointed if you quit but there are other people available. Whatever you decide is up to you."

That was it, out he stormed along with Barb. Shirley went off to bed and Rod and I turned the television on. The rugby international was playing. Australia, against all odds, won the test match. Rod and I were elated. The era of Australian rugby dominance had long gone and we had deflected many a taunt from our mates in other countries. A rugby victory was a better note to go to bed on than a dispute with Ken.

When I got to the ute, I saw that Ken had taken a tarp off the awning and covered the back so that I could sleep comfortably. In the morning I thanked him for the tarp.

Barb said she had been frightened during the night because she had heard someone walking on the stones. Ken agreed, "Yes it is frightening." Rod had gone to the toilet at 2.00am. Thinking about the night's proceedings had kept me awake so I had seen him. I agreed. Rod in his pyjamas is a scary sight.

Rod and Shirley left for Brisbane. I left for Hebel down the road and made 35 kilometres before Ken collected me and brought the kayak and me back to Dirranbandi for the last night there. Nothing was mentioned about the discussion the previous night. I put it out of my mind. My little toe was worrying me but there was nothing to be done except to put a fresh Band-Aid on it.

July 1st was a day for some reflection. Exactly one year ago I was close to death, lying on the Plenty Highway, bloody froth coming out of my mouth. After that, everything was a lucky break. The police station was close; there was a 4WD ambulance close; there was an airstrip nearby, and there was an air ambulance to pick me up and get me to Alice Springs Hospital. There was also a paramedic, Davo, on the scene. Two months after that I could sleep lying down. Six months after that I paddled 50 kilometres. One year after that and there were more than 800 kilometres behind me on an adventure that was unfolding into a fantastic learning experience. It was a day to be savoured. No one could know how special it was to be alive and to be here. The moment was very personal.

The next day there were quite a few cattle grids. On each side is a sheet of white tin that delineates the grid. Beside one grid there is a half a tank on its side and under that a chain and a dug out area. Obviously a farmer ties his dog up here and the dog digs until he finds cool dirt. There were some blokes building a fence about 10 kilometres out of Hebel and that was it for my day's excitement until I got to town.

Perhaps 'town' is an overstatement. There is a café which gets passing trade, mainly from buses. There is a general store which has accommodation, the caravan park, and a restaurant – yes a restaurant – attached. Across the road is the Hebel pub and diagonally opposite that are the cabins for the pub accommodation. There is a hall to the east of the pub, plus a school and a house around the corner to the west, but the western part is not evident from the centre. That's it, the town in its totality.

The pub looked very inviting. It is one of the world's great pubs. The bar slopes down and most things look just a bit crooked. John Murray, a famous artist from Lightning Ridge 70 kilometres away, has painted his emus and witty signs all over the place. What better place to slake one's thirst. Chris, the publican, had the St George paper behind the bar so as well as knowing all about the golden oldies game, he knew we were coming his way.

In John Murray's gallery is a painting for sale at $38,000. He is not cheap. How did Chris afford the artwork around his pub? His explanation is a gem: "John is a good bloke, and he doesn't mind a beer." People look after each other in the bush.

Unfortunately Chris got to know me a bit too well that day. He never has a beer until 5.00pm but he does serve plenty. Before long the twenty dollars on the bar had evaporated and a top up was needed. It does go quickly with beers for Ken and I plus white wine for Barb. After about six, Ken and Barb went to prepare dinner but I stayed on for another hour. Chris says he knows how many beers I drank but I have no idea. My feet did not hurt walking home and I slept soundly.

Chris is interested in hydrogen as a fuel and has managed to make a motor run without blowing himself up. This may be good luck rather than good management as he says it does make a good 'pop'! His story needed to be on camera so we had arranged to film the following night. Filming went fine until I touched on the tricky topic of suicide. It had been raised way back in Ipswich by the truck driver who had seen a lot of rural suicides in Victoria. There is no issue more serious than life and death and what better person to ask about deep and meaningful stuff than a publican.

The response was unexpected. It's not just landowners who are at risk. The businesses that are as dependent on them are just as vulnerable. This was the very point that Robert Buchan, mayor of St George, had stressed. While he spoke about the issue Chris rapidly became emotional. The topic had a deep effect. He took a step backwards and Ken turned the camera off. 'Shit,' I thought. 'This is powerful stuff and we need to get it, even if we don't want to use it.'

I tried to turn the camera back on but had forgotten what all the buttons did and just succeeded in changing all the settings. The moment was lost, gone forever. 'So much for his insistence in Toowoomba about not turning the camera off until we both agreed,' I thought. Having been through a bout of depression myself, I believe that if you have not decided the method you will use to suicide, then you still have some way to go before it is likely. But that is just my personal opinion.

The pub was packed with people on the second night while they watched a very tough State of Origin rugby league match. We had eaten dinner at the restaurant across the road but that night I also put away two of Chris's steak and mushroom pies. Physiologically something had happened. Until now my food intake was not particularly high and I had gradually lost weight. At the time it was not evident that there had been a major shift in my body requirements but looking back, it is obvious that I had used up my fat reserves and needed to eat more every day.

Hebel to Goodooga is a gravel road some of the way, but much of it is just dirt. The border crossing is a cattle grid with the signs welcoming you to New South Wales shot to bits. The shooters cluster their shots until a piece about the size of a beer glass gets blown out. Ken and Barb arrived in time to film the crossing. This coincided with an important interview with ABC Radio which was linking comments from people in Queensland, New South Wales and Victoria. The interviewer just asked me to describe the scene and what I had witnessed since Toowoomba. Phone reception was patchy so I drove the ute 200 metres until the signal was good. Ken and Barb took their chairs and sat under a tree. There was no way I could involve them in the many radio interviews that were conducted on the phone so they had lost interest in them. It was also school holidays so they did not have the engagement with the school presentations either.

The following day I lost the signed safety vest. This was a major blow. It was something I wanted to keep forever. Taking my jacket off, I hadn't thought about the vest. It had blown away without me realising it. Despite frantic searches along the road, it was gone. Ken and Barb spent the day in Lightning Ridge. I passed through Goodooga. On the last day before they left for their break in Sydney I fortuitously stopped at a floodway. This made it easy to remember where to come back to.

It was Thursday 5th July and Carol was driving out to Moree to meet us. Ken and Barb were to take her Astra to Sydney, attend a wedding and Carol would join me on the road for five days until they returned. First, I was heading home with Carol for a bit of rest, recreation and recuperation. Specifically I needed some dental work, the ute needed a windscreen and the kayak badly needed its worn and broken wheels tidied up.

Chapter Ten

We packed up and I drove with Ken sitting in the back seat. At Collarenebri, which is on the Barwon River, we stopped for diesel. With the kayaks and stickers the unit is very conspicuous and the service station proprietor quizzed us about what we were doing. He bemoaned the damage that cotton farming had caused to the river. I said, "Surely it brings wealth to the area so that guys like you have a good business." "But at what cost?" was his response. Touché.

Carol was waiting at Moree. We took all of Ken and Barb's gear and loaded it into her car. When we were about to head our separate ways, Carol noticed a pair of gardening gloves where Barb had been sitting in the ute. She pointed them out but Barb said not to worry she wouldn't need them in Sydney. They had the Apple computer and the external hard drive to take footage to one of their daughters who works for Channel 9.

Back home at Karalee, I went to the dentist to have two teeth filled. They should have been crowned but there was no time. Luckily there was no pain but they had fallen apart in dramatic fashion a week earlier. The new windscreen was fitted and John Crocker helped me sort out the wheel bearings for the kayak wheels as well as rebuilding the front wheel steering bushes. John had originally built the wheel arrangement. He and I make a good team and the design was about as good as it is possible to get. John has always been an integral part of the crew. It was a very busy day.

Some of the support team were meeting at Rod and Shirley's on Sunday afternoon to hear something of the trip. Not many could come because it was only arranged at the last minute when I had decided to come back to get my teeth fixed and fit a new windscreen.

Late that afternoon, Friday, Ken called from Sydney.

"Your little outburst at Dirranbandi has cost you dearly. Barb and I have lost all respect for you and have decided not to continue with the trip."

I replied that I was sorry they felt that way but if they had a change of heart to let me know. Ken said that they would return Carol's car to Karalee on Thursday.

Carol was shocked. So was I but more than anything else, I felt a huge relief. We had experienced much bigger and more serious shocks in business. It was time for action.

Ken sent the following email to the members of the support team:

> *Unfortunately Barb and I will have to resign from our positions as support crew for K4e.*
>
> *We have been disappointed on a number of occasions with Steve's attitude towards us, other people and the integrity of the trip.*
>
> *We lost all respect for him when he embarrassed and belittled us in front of visitors (Rod and Shirley) saying we were not doing the job. We were only "a fair weather support crew".*
>
> *Barb and I have worked bloody hard from daylight to dark to ensure the success of the trip and Steve's comfort and safety. We have done our very best but it is obviously not good enough.*
>
> *We are very sorry that things have come to this as we were enjoying the experience of meeting new people and filming the trip but we will not be bullied, especially after the significant sacrifices we have made.*
>
> *Steve told us during our "discussion" that he had other people available to take our place. We sincerely hope this to be true so the project can continue.*
>
> *Barb and I would like to thank the many people who have been so kind, friendly and supportive to us on this journey.*
>
> *Sincerely,*
> *Ken McLam*

First job was to draft a response. The following email went to the team plus to Ken almost immediately.

> *Dear all,*
> *I have accepted what Ken has said. The day before had been the toughest physically by far, much worse than the Toowoomba Range, so I guess I have found out something about myself under extreme physical conditions.*
>
> *Obviously I feel there are two sides to any story. The problem is going to be the short term ie getting people to change plans straight*

away. Not sure what plan B is going to be in the next two days but I will sort something out. Right now I will just have a think and explore the options. What we have learned is that it is a great chance to meet lots of different people and learn all sorts of interesting things about our country.

The shame is that Ken was learning more and more about the photography and even the professionals for the 7.30 Report said that his footage was good. That expertise cannot be replaced. There is no doubt that I greatly appreciate Ken and Barb's support to date and wish them well. There are no hard feelings on my part just a sadness that it didn't get to the finish.

Steve

Second job was to get on the phone and sort out a new plan. The plan had been that Carol and I would be back out on the road on Monday and that Ken and Barb would return with her car to wherever we got to. We would continue the trip together, probably to near Brewarrina, and Carol would return home. Now that her car was coming back to Karalee that plan was down the drain.

After an hour on the phone I had some hope and some disappointment. Finding a crew within two days was tough. It was a big thing to ask of anyone. There were many people who had expressed an interest in being part of the crew but I had turned them down earlier because Ken and Barb had taken that role. Now those people had structured their lives in other ways.

That night our son Jonathan said he would do the job. This was exciting to contemplate but it was a big commitment on his part. Carol and I asked him to think about it some more. He was due to fly to Canada on a working holiday in mid-September.

The following morning there was an email from my Hash House Harrier mate Monty. Monty had almost died when his motor bike hit emus two years before and he was the one responsible for contacting the hash guys and getting Whale to turn up at the West End boat ramp. He had seen Helga, a fellow hashman at a party the night before. Helga was working at his parents' backpackers' hostel but tentatively casting about for an engineering job. He was available immediately and keen for a bit of adventure for a couple of weeks.

Monty's name comes from Montgomery – he drinks like he has just come out of the desert. Helga is relatively new to the Hash House Harriers. His name derives from some Swedish girls he organised to serve beer at his parents' backpackers where the Brisbane Hash Harriers meet. His real name is Dave Ryan.

I called Helga/Dave, who seemed to be still operating under the effects of the previous night's party. He was keen to come after we discussed what was involved. The following day, Sunday, we arranged for him to ride his motor bike out to Karalee at 6.00am Monday morning. He would be available for two weeks but was flexible if a little longer was required.

Jonathan had decided on his plan. He needed two weeks to get his affairs in order and wanted to come on the journey after that. Carol and I both agreed that this was an excellent arrangement. The plans were made. He would replace Dave in about two weeks.

Sunday at Rod's loomed and I was apprehensive about everyone's reaction to the recent dramas. When I got there most of the team had turned up. They all know me quite well and wanted to show their support. I was very touched. Eventually, it was time to broach the topic of what had happened. I explained the situation as best as I could and was very pleased that Rod and Shirley had been there at the crucial time. Shirley didn't speak as I had, but she let it be known to most people that she thought the issue was that Barb was simply not up to such a trip.

It was cold and dark at 6.00am on Monday morning. What if Dave didn't show? A motorbike could be heard in the distance but it went past to the end of the street. The house lights were on but I had been too slow getting out to the road. My phone rang, "Dave here, where's your place?" I told him to come back 200 metres and I would be on the road.

Chapter Ten

In he came, large pack on his back and another one on the front. The bike was a 250cc screamer – a pocket rocket. Jonathan had owned one like it but he had decided bikes were not a good idea after he slid past me on his back at 120km/hr. Carol made a coffee to warm Dave up. We loaded his gear into the ute and then we were off. It was less than sixty hours since Ken's bombshell call and plan B was in operation. Total time to come back, effect repairs to vehicle, kayak and body was three days plus a day each way travelling. Not a bad effort!

Part Two

There are some great people keeping an eye out for Steve ... he is well remembered and the yarns abound. As Banjo once wrote, "He remains undaunted and his courage fiery hot."

David Bristow, CEO Simmonds & Bristow

Maps for this part appear on
page 108 and 180.

CHAPTER ELEVEN

The Bokhara Delta

DAVE IS AN ENVIRONMENTAL ENGINEER. On the trip out to Goodooga he marvelled at the huge farm dam walls (ring tanks), commenting that hydraulic modelling would be useless without detailed knowledge of all of these structures. He was also able to point out where farmers had planted trees for wildlife areas. His perspective was slightly different from mine, so we learned from each other.

It had been a great break. The new windscreen made the vehicle feel like new; the boxes and the back of the ute were reorganised and we had repaired all equipment. I had been able to wish my father an early happy birthday for his 80th which I was going to miss on 21st July.

We stopped at the Hebel pub to give Dave the experience of a classic outback pub. Chris, the publican, had let people know about the signed safety vest I'd lost but it hadn't been found. The route out had covered a lot of what I had walked so Dave gained an appreciation of what was involved physically. Talking to Chris he was introduced to the friendly culture of the bush. We downed our beers and headed across the border and down the dirt to Goodooga.

Goodooga was our first New South Wales country town. Was it a coincidence that it was the first town that looked like it's dying? There is not even a general store. It closed long ago. So did the bowling club. 'Mismanagement,' the locals say. There is a post office selling ice

Chapter Eleven

creams and drinks, a police station, a school and a small hospital. Unlike Hebel there are enough streets to make it look like a small town. A bitumen road connects to the Castlereagh Highway but all other roads are dirt.

By the time we arrived at the floodway where I had finished the previous Wednesday, we were both immersed in the flat, arid isolation – away from the city and all it stood for. There were plenty of trees by the floodway so we pitched camp, lit the fire, and planned the next few days under pleasant circumstances. Dave slept in the back of the ute and I contented myself with being under the awning off the trailer.

We were just east of the Bokhara River which had no water in it, and west of the Narran River which had no water either. Brewarrina was well over 100 kilometres to the southwest but we expected to find water in the Barwon River there. Walking down the road with the early morning sun just over the horizon, it felt great to be under way again. For the sake of making maximum progress it was an advantage that no internet coverage was available until Brewarrina. That meant no daily reports were needed for Geoff to put on the website. Traffic volume was about six vehicles per day which is good for company but quiet enough to enjoy the isolation.

Back in Brisbane we had been shopping for some more shoes. They were still cheap joggers but they fitted my feet. Normally my feet are size nine but after a few weeks they became swollen. Cramming seriously blistered size ten feet into size nine shoes is not fun. The first few steps are awful. With the worst of the blisters gone and shoes that were the same size as my feet, walking was a lot less painful than it had been a week before.

On long journeys things happen slowly. It is necessary to think rationally and be observant. I hadn't noticed the first flat tyre back on the road into St George even though it had been deflating all morning. I just thought I was getting tired. Sometime later, I noticed that the harness frame had gradually slipped down, putting more load onto my hips. It wasn't until I stopped and bent the aluminium bars back up that I realised it had happened over quite a few days. When fixed, the difference was amazing. Now, I was going through the same experience with my feet. The increase to size ten had been so gradual that it was not until I tried the larger size shoes on that I recognized the problem.

Dave drove past after about an hour. He had packed up the camp and was now focused on the day ahead. We had already agreed to aim for a 40 kilometre day even though it was a dirt road and therefore harder work than a bitumen one. Being a member of the Hash House Harriers he took a classic runner's approach to the problem. His plan was to set intermediate goals, time me over these distances and provide feedback to motivate faster times. First up was 15 kilometres to be completed in three hours and then 15 minutes' break. His philosophy was different but it seemed reasonable to me. With the first goal down it was on to the next.

Dave had marked the spot where we had morning tea, driven five kilometres ahead, parked the ute and trailer, run back to the spot and then made sure he got back to the ute first. This turned into a race. One party had to drive five kilometres and run 10 while the other had to drag a kayak for five kilometres Do this eight times and the fella with the kayak has finished his 40 kilometres.

The reality was not quite that easy, as it takes an hour to pack up the camp and a similar amount of time to set it up again by the time Dave got the wood and lit the fire. Interviews for the video were also required. Dave did manage two sets of runs most days, giving him 20 kilometres. Occasionally he did two 15 kilometre runs. On that first day we stopped to film so to get my 40 kilometres completed, it would be necessary to finish an hour after sunset but the road was wide, flat and should be easy to see.

The countryside improved as we travelled south west. Some grass appeared in the paddocks. Beside the road there were green patches of grass and long stretches of the table drain beside the road held water. There had been 50 millimetres or so of rain through the area just a few weeks ago. With low temperatures much of the water had remained.

Chapter Eleven

I was excited again. Dave was keen, Jonathan would be joining me in two weeks and I no longer had to worry about the commercial aspects of filming. That had been an added stress, constantly thinking about helping Ken to get good footage. Dave was not trained to use the video camera and neither was Jonathan so a saleable product was unlikely, even though we would try.

During the first six weeks I hadn't confided in Carol about the strains in the team. She had her own problems back home. Now she had seen it all blow up and was caught up in it. Ken had rung her to say they would leave home early Thursday morning to return the car. She felt that he was fishing for thanks even though they had kept her car for six days after resigning, leaving her seven kilometres from the nearest shop without a vehicle from Monday, when Dave and I left, to Thursday when they arrived. The Apple computer and the external hard drive with all still images and edited video footage were also with Ken and Barb.

It was as if the border between Queensland and New South Wales was the decreed cut-off point for rain. Over near the coast on the Clarence River it was the same. New South Wales had received rain while Queensland stayed dry and parched. Tony Schneider, a farmer in the area, had stopped for the usual country chat back near Hebel, saying that we'd notice the country getting better as we went south. At the time he was delivering feed up to Roma which was locked in drought but invited us to stop at his place when we got there.

Nearing his property we called him on the radio and arranged to meet at his farmhouse. Because there had been good rain he had just planted a crop of wheat but still had time to show us around. At our roadside meeting he had explained that his last sorghum crop had been in 1990. It was becoming increasingly difficult to fit it into the short winters. Late planting and early hot temperatures were burning the crop off before it was due for harvesting so they had given up. Since then the chance of damaging high temperatures coming early had increased.

Tony is passionate about his environment. He calls it marginal country but says it is profitable if you understand it. His small weir across the

Bokhara River contained water. Beside the water were healthy coolibah trees and lignum, a type of bush that cattle will eat. Away from the water the coolibah trees looked like they were dying. Tony said they wouldn't all die but the landscape was changing. In 1922 only 250 millimetres of rainfall was recorded, making it a serious drought year. But there were four overland flows that year. Despite the drought, there was enough water for the trees and grass to survive quite well. "We don't get those flows anymore," he said.

"Without flows to wet their feet, why won't the coolibahs die?" I asked.

"You see those flat bits with nothing on them," he explained. "Well they are a natural part of the landscape and act like a roof does for the house water. They catch the water and run it off to the other areas that need it. That gives the trees near these areas enough to survive from rainfall. As you can see they do not look healthy like the ones near the river do, but they will survive. These areas are called claypans and they are an important part of the landscape out here."

The claypans certainly are noticeable, being up to a kilometre square. Is he right about the coolibahs being dependent on them? Who knows, but clay pans just don't seem right. Why are there areas where things grow fine and beside them there are areas where absolutely nothing will grow on the hard, baked soil?

The weir is tiny. In fact, the river is tiny. One waterhole back up on the Balonne would have dwarfed this reservoir. Some people would call the water storage an overgrown puddle, but these guys are frugal with their water use and it's a very valuable asset. We had heard much of drought and seen many drought affected areas but rainfall in the first half of the year here had reached 400 millimetres. Average rainfall is only about 400 millimetres per year so this was a good year and Tony was very pleased. Farming is very much a lottery. What he had experienced was not widespread.

Tony collects rainfall records to understand cycles. He believes that the St George irrigation area is overdeveloped. Before 1950, records show times much dryer than the period 1950 to 1980. It was these later years that were used to allocate water. Now, the climate seems to be returning to the pre-1950 rainfall. "St George is advertising the potential for major extensions to irrigation. They just don't get it," he says. As for Cubbie Station, "Cubbie is part of the problem, but only a part."

He says, "Areas of mitchell grass are now gone. It's more difficult to get water up from the Artesian Basin. Things are changing. It is much harder to make a dollar today. Mind you, the wheat crop that's in now will be worth a million dollars when it's harvested. That can carry you through."

Back on the road we had covered 30 kilometres so Dave left to set up camp 10 kilometres ahead. An hour later two Aboriginal blokes drove up in an old truck.

"That your mate bogged up the road there?" they asked.

Funny question I thought. "Is he all right?"

"Yeah, he reckoned he was fine. Didn't want any help." With that they drove off saying they would see us on their way back down from Goodooga the following morning.

Eventually Dave drove towards me. Stopping the vehicle he said that 38.5 kilometres would have to do so we pulled off the road and set up camp. It was a very ordinary place to spend the night. If there was a tree hiding somewhere neither of us could see it. We were on a very flat bit of ground without even salt bush to break the wind. The temperature dropped to zero, and there was nothing whatsoever to burn to keep warm. While cooking dinner on the fuel stove Dave confessed to his adventures. Trying to execute a U-turn, he had driven one side of the ute into the mud and water in the table drain. The ute is 4WD but is not fitted with limited slip differentials so if one side goes down, both the front and back wheels spin and the vehicle stops. Dave had to dig his way out by excavating a track and filling it with dry soil. It took him about an hour to do this but it was a valuable lesson.

Dave slept in the back of the ute again and was snoring by 7.00 o'clock. I made it to my usual 8.00 o'clock drop off time and I enjoyed the stars until then. We might be on a treeless plain with temperatures hovering around zero but the night sky was brilliant. To enjoy its beauty I was sleeping under the awning arranged so that the dew would stay off the swag but I could see more than half the night sky. Poets, musicians, artists, people much more sensitive than we engineers, delight in describing such images. It was a comforting feeling to just lie back and take in the clarity of billions of bright pinpricks of light above me.

We awoke to the desolate scene and I was keen to get going. Dave was enjoying his adventure and thought treeless ground, cold and exposure were just part of the fun. Like yesterday, he drove ahead five kilometres after marking where we were by drawing his foot though the dirt. It certainly is weird to be out there, walking along towing a kayak and watching someone running towards you. It is an extraordinary sight watching a lone runner disappearing into a treeless, stark plain.

Dave's pacing me was working. I was making the distance in good spirits and enjoying being partnered with someone who understood the physical aspects of the challenge. I looked forward to Dave running past in either direction, the camaraderie of the race and the fact that he was doing all this to help me. The difference in the atmosphere was palpable. I looked forward to waking up in the morning and the stresses and strains of the physical journey didn't weigh on me as they had done.

Just to cap off the day, the dirt road had given way to bitumen, making pulling the kayak much easier and putting less strain on my feet. The bends in the road were still few and far between and more than a kilometre long. The country was still flat but the bitumen at least gave the appearance that civilisation was not too far away.

Downstream of Tony's place is Bokhara Hutz. It is a farm run by Graham and Cathy Finlayson. At just 18,000 acres it is considered by most people too small to be viable. Without much money to back them, it is all Graham and Cathy can afford but their operation is something to see.

Dave had contacted them by radio in the morning and they had come out to the road to meet us. Two were on horseback – Lotte the Dutch visitor and Harriet the eldest daughter. Cathy drove their ute with the two youngest, Lily and Kate, and Graham walked with me. The horses were very tame so despite my apprehension they were not troubled by the kayak. Graham had cooked the lightest scones I have ever tasted. The recipe has something to do with lemonade but the details are confidential.

On reaching the entrance to the farm we agreed to keep going until 4.30pm and then come back in the ute. Dave drove into Brewarrina to collect liquid supplies. The Finlaysons and Lotte put the horses into the float and went into the farm. I pushed on and eventually Dave arrived back with beer and a bottle of rum.

The property actually covers two rivers, the Bokhara which flows south to the Barwon, and the Birrie which heads west to join the Culgoa. Both rivers start at the Balonne Minor before it branches out. The area is a huge inland delta. Its history is of overland flows from the big Queensland wets but these floods are now few and far between. Graham had been trying to manage a property without the historic life giving water over the land.

Until 1999 Graham and Cathy were part of a family partnership. They bought out of that to try their own methods. Normally one sells out of property, not buys out of it so that was a stark reminder of how difficult life can be out here. The property is in cattle and sheep country. Unlike Tony upstream, they have no irrigation licence. One good year was all they got. In 2000 the worst drought on record enveloped them. It lasted six years and halved the average stocking rate.

With a big mortgage and little rain, things looked grim. Many seek handouts under such conditions. Many curse their bad luck. Most stoically survive. Graham saw this as an opportunity. He thought that if he could develop farming methods that would sustain them through such times they would indeed be viable long term. They changed their philosophy. "You can't drought proof the landscape," says Graham, "so we tried to drought proof the business."

They read *Natural Capitalism* and studied the principles. They believe that owning land is not a right, it is a privilege that carries with it the responsibility to look after it for future generations. They want the world to look at everything holistically, to take into account environmental and social costs as well as economics. "Imagine what Brewarrina would be, imagine the community we would have, if small properties like ours could support a family. It would be a thriving family district again like it was 40 or 50 years ago. I do not agree with the idea that you have to get bigger or get out," said Graham.

Now the farm is a fully certified organic property. The business does not own any livestock. It provides agistment for organic producers. It is part of a group called 'West 2000 Plus' which is comprised of ten properties. Some have cultural responsibilities, some fauna responsibilities, and three properties including Bokhara Hutz have ground cover responsibilities. These three must maintain or improve pasture even in the driest years. Graham had this to say:

> This is semi-arid, brittle country, just like many of the rangelands in the world. It seems to me that it is no accident that places like the Middle East are now desert but were much more fertile thousands of years ago. Australia is heading the same way unless we change our ways. Grazing techniques dating back to European practices are not appropriate.
>
> If I blame changing climate, weeds, stock prices, or anything else out of my control I will never change. But I can influence my decisions such as when to put stock on, when to take them off, how many. I concentrate on our land, trying not to worry about the things about us that I see are wrong. A free enterprise market is fine up to a point but you should not be able to externalize costs. If you impact negatively on anything, be it a neighbour, the environment or society, then you should suffer a cost penalty.
>
> We have closed off most of our stock dams. This has controlled the kangaroos and allowed us to move the stock where we wanted. Once scalded country that was used as catch areas for stock dams is now starting to get some growth. Thirty kilometres of poly pipe now transfer water around the 35 paddocks. Water comes from the river. There is just enough water running to the river from rain to do the job without relying on flows from upstream. Recently we were lucky enough to be allowed access to an artesian bore.
>
> The word 'sustainable' has been hijacked. To me it is only sustainable if you can go on doing it indefinitely.

I liked this bloke. My message to the school children was always "If it is sustainable you can keep doing it forever."

Graham's opinion of claypans was exactly the opposite of Tony's. Tony thinks they are natural. Graham doesn't. Tony wants to keep his. Graham wants to revegetate his. It would be almost another thousand kilometres until a 67 year old farmer who had lived in the same place all his life and flown light planes all around his area, caused me to form a strong opinion. At this stage, Graham's explanations just seemed more logical than Tony's.

Like Tony, Graham had rainfall records for the past 130 years. His story was that there were two overtoppings in 1922. Whether his figure of two, or Tony's of four, is correct is irrelevant. The fact is that in a drought year the local farmers still benefited from what we would call floods. These provided valuable, deep moisture and allowed the trees to survive. Many people tell Graham that the land initially had no trees. This is a common story. "This country is not supposed to have trees on it. There were none before white man came and now we are not allowed to chop them down," can be heard all over the country, at least from St George to Brewarrina. "If this is true," says, Graham, "then why do trees start to shoot when you take the cattle off the land?"

Graham's goal is to double what is thought to be the stocking rate for his land. In 2005 he won the New South Wales 'Young Farmer of the Year' award. Many people have a stake in his success, they wish him well, and it is hard not to be infected by his intelligent optimism.

Chapter Twelve

Brewarrina – Bourke

It was just over 30 kilometres into Brewarrina – an easy day. My right foot had been sore the day before, maybe as a result of going hard between Goodooga and Tony's the day before that. Dave's goal setting certainly livened up the day and racing him was fun but I began to wonder if the constant pressure might be counterproductive in an ultra long distance event like this one. Dave ran 15 kilometres in the morning. This staggered me. He and Graham had demolished a bottle and a half of rum. Hash men spend their lives training in both skills – running and drinking. The training seems to work.

Brewarrina boasts a modern and attractive hospital. Graham's sister, Heather, works there as a nurse practitioner. This relatively new job classification allows nurses to take over some of the doctors' roles. Heather and her husband have a property back up the road near where Dave and I had camped on the treeless plain, but they also have a house in town. Heather stays there when she's working, rather than drive the 60 kilometres back to the property. Graham and Cathy had arranged for us to stay the night with her.

About 20 kilometres before Brewarrina, there is a monument near a sign that reads 'Hospital Creek'. The 'creek' is marked by a bridge over a slight depression in the ground. The word creek seems incongruous, particularly as there are no trees. Perhaps a long time ago it was a real creek with water and billabongs, maybe swamps off to the side. Perhaps

CHAPTER TWELVE

it is similar to Bill's filled-in creek way back near Toowoomba and the changes have been enormous. The monument is about 20 metres off the road. Small painted stones edge a boomerang-shaped path. At the top of the bend in the boomerang, ie the path, the monument itself stands about one metre high. To get into the area you have to climb over a barbed wire fence. I wondered what it all meant. Someone in town would know.

Coming into Brewarrina the road passes over the Barwon River. Yes! There was water in it. What's more, the water was almost clear. Had it not been less than ten degrees with a cold wind blowing it would have been very inviting. There are houses, bitumen streets, and it looks like a normal small town. Rounding the corner and heading for the Council-operated information centre, the scene changes. Like many towns in the area it is a shock to behold. It looks like it is under siege. Iron bars and shutters cover everything. Groups of kids hang around the streets. Most people drive through without stopping. Many of the grey nomads say they are frightened of the place.

A young bloke was walking towards me. "G'day mate," I said. "Hi Bro," was the response. This was repeated many times. If you say hello to the kids they are friendly and responsive.

The information centre is run by Fran Carter who had contacted us by email. The centre is well worth a visit as it provides a wealth of information about the local area – particularly the Aboriginal fish traps. Fran can tell you anything you want to know including information on the monument by Hospital Creek. It was built in memory of the Hospital Creek massacre that occurred in 1860. During construction of the monument, one of the Aboriginal men died suddenly and traditional law prevented the Aborigines involved from working any further on the project.

It wasn't long before we found out that water issues here are as difficult and divisive as anywhere we had seen. The town weir once had a section with removable boards in concrete slots. Boards could be removed to lower the weir level and let water downstream if required. The timbers kept being removed. The Brewarrina people reckon it was the 'bastards from downstream' at Bourke. Eventually the opening was filled with concrete and it is now not possible to lower the water level below the weir crest.

We talked about the town, the kids, the troubles. One of the kids is called Cyril. The round steel cages that the Council uses to protect new trees disappear and are found in the river, converted into fish traps. Cyril gets the blame for this. In fact he gets the blame for most of the misdemeanours that occur. "Look out for him," Fran said, "he has too much time on his hands. Because he's a likeable kid people have taught him too many tricks."

She had arranged for us to meet some of the community members outside the information centre. Moses, a very round man with a long beard, is the town encyclopaedia. There was talk about him being on Sale of the Century, a well known national television quiz show. Moses said it was 277 kilometres by river to Bourke and that river distances are about three times that of road distances.

The town weir is close to the information centre. I walked the kayak to it and launched through reeds just above the concrete wall on the edge. The flow was about 400 megalitres per day which, coincidentally, is about the same as Brisbane uses. The dilapidated fish ladder had some timber cross members still intact which provided perfect kayak slides. I simply held onto the back of the kayak and allowed it to slide down the timbers into the water below the weir among the fish traps. Dave filmed some of this. All that is visible of the fish traps through the plentiful reeds are the tops of large semicircular rock walls. These walls had been constructed by the Aborigines to herd and trap fish but hadn't been used since white men had arrived. There are some who say that this is the oldest man-made structure in the world. Getting the six metre kayak through the fish traps was a challenge in its own right.

Downstream of the traps the river becomes a large waterhole without discernible flow. There were a couple of kids close to me and a much bigger group further on. "Can I have a go in your boat, mista?" said the older one, about twelve. "I've never been in a boat."

CHAPTER TWELVE

Dave was fetching the ute to take the kayak back to Heather's for the night and we didn't need to cover any more distance so I agreed. "OK you can take it down to the big rock and then you have to turn it around and come back."

"Can me mate sit on the front?"
"If you get down to the rock and back easily then maybe he can. What's your name?"
"Pundoo."

So I climbed out, instructed Pundoo in some basic arts of paddling and away he went. His balance was perfect but his technique was what would be expected from someone who had never been in a 'boat' before. After lots of instruction from the bank I decided that his little mate was not allowed on the front and told him so when he got back. Pundoo didn't like this assessment and told his mate to hop on anyway but I grabbed the kayak. His mate's attempt was pretty feeble so I knew his heart was not in it.

Walking back up the bank to meet Dave, Pundoo asked me for money.

"What do you want money for?"
"Got none. Need money for bait. Can't catch fish without bait."
"Well I don't need money on the river so I don't carry any, sorry."
"Not even a dollar?"
"Nope."
"C'mon Cyril, let's go," said his mate.
"Is your name Cyril?" I asked, Fran's briefing flooding to mind.
"You sure ya got no money? I need money for bait."
"Like I said, Cyril, I don't carry any because there is nothing to buy on the river."
"OK. See ya. Me name's not Cyril. I was joshin' ya."

Fran later confirmed that I had met Cyril. There had been some spray can graffiti in town with the name Pundoo on it. Cyril denied it was him. Despite his reputation, I liked the guy. School does not interest him so he doesn't go. He's obviously a smart kid with too much time on his hands and too many temptations. He is bound to get up to mischief.

We had been told which house was Heather's and found it easily. When we pulled up, all the doors were open but there was no one home. Finally Heather arrived, explaining that she had been caught up visiting friends, and instructing us to put the ute and the trailer in the back yard. The house was very old – in the city it would be treasured and renovated to highlight the old charm, but out here it was just a house.

I rang Carol to see how her day had unfolded. Ken and Barb had returned her car before lunch. Not a lot was said between them. Carol gave them some fold up chairs that Ken had bought with the donation from BCF. Although very comfortable, the chairs were too big so we had bought some small ones that flat fitted neatly into a corner of the trailer. Ken gave her the Apple computer but not the external hard drive. Now that she had the car, Carol drove to Tamworth to collect her mother and father to bring them out to Brewarrina for a visit. She brought the computer with her so Dave could continue the role of filming and making DVDs to send to the media.

We arranged to meet the next day at 4.00pm at the information centre. Carol was very pleased to have wheels again. Ken and Barb had kept her car six days after resigning. They hadn't left one thing of theirs in the trailer or the 4WD except the cheap gloves Carol had pointed out to Barb the day they departed. Even the reflector for filming had gone. Obviously they took Carol's car away for a week, knowing full well they would not be back. Not only was their behaviour incredibly selfish and rude but it seemed designed to cause maximum trouble for the trip. Had I been so bad? At least it was over and done with and we now had a team that worked.

We bought Chinese food from the RSL club and watched the 7.30 Report coverage of our expedition. It was very well done. After that, the ABC screened the British documentary, *The Great Climate Swindle*, made specifically to counter Al Gore's award winning *An Inconvenient Truth*. Tony Jones, the ABC presenter introducing the film, pointed out some fatal flaws in the documentary but was very aggressive in his arguments. His points were factually correct but I thought the level of emotion Tony displayed was likely to have a negative effect on the audience. Scepticism is healthy and an essential part of science. To deny the role of scepticism in science is not healthy and not productive. Because many climate sceptics are not very logical, allowing the argument to get emotional only works to their advantage. Science and logic must be paramount in all of the debate. Emotions need to be kept out of it.

The next morning the John Laws radio show discussed the documentary and Bob Carter was the guest, supposedly to provide a balanced view. Bob's not a climatologist, he's a palaeontologist, stratigrapher, marine geologist and environmental scientist. It's like one of my customers getting a balanced opinion about my gate designs from a process engineer. The gates may go into sewage treatment plants that a process engineer designs and operates but he is not qualified in the details of gate design. The radio show, television documentary and ensuing debate all disturbed me. When I set out from Brisbane I thought the job of explaining climate change was done, that people understood the science and accepted it. I was finding out just how wrong I was. To understand why people do not accept the science or think that the scientists do not agree all you have to do is look at the history of the tobacco and Star Wars debates in America.

The George C Marshall Institute was founded in 1984 to promote Ronald Reagan's Strategic Defence Initiative (SDI, colloquially called Star Wars Technology). The three scientists paid by the Institute set about discrediting 6,500 physicists who didn't agree with the SDI program. They successfully argued that 'balanced' media reporting meant that just over 50 percent of media articles should support their point of view. The ratio of 6,500:3 scientific opinions became 50:50 to anyone who relied on the media for information. These techniques had been developed and perfected earlier by the pro-tobacco lobby. Some time after Reagan dropped SDI, these guys switched their attention to global warming, using the same tactics and many of the same people.

In Australia there is a fossil fuel lobby referred to by Clive Hamilton in his book *Scorcher* as the 'Greenhouse Mafia'. This group was extremely influential with the Howard Government. There are also four very prominent contrarians who argue against the science of greenhouse gases. One of these is Bob Carter. None of these contrarians has to my knowledge had a paper on climate science subjected to the scientific rigour of peer review.

The science on global warming is unequivocal. There is no doubt that warming is occurring and it is caused by man's influences. The slide presented by Roger Stone at Toowoomba was a very powerful demonstration of this and quantifies the actual effect of mankind on the climate.

Dave and I did some filming around Brewarrina, fixed punctures and finally headed down river. The flow had dropped a lot and was just enough to float the kayak between waterholes. There were a few tree guards in the river which had no doubt been used as fish traps. There were some more stone circles which were the remains of old fish traps. Unfortunately for me there were a lot of logs. Logs are good for fish as they provide habitat for them to live and breed in. Logs are not good for kayaks. Kayaks have to be manhandled over them, sometimes two metres out of the water. The late start, logs, shallow water and 4.00pm date to meet Carol meant that we only achieved a little over 20 kilometres for the day; but it was on water. There would be no more walking for a long way. My feet were pleased.

Back in town we checked into the Middle, an old pub. There were once two hotels in town but this one was being converted to accommodation and a restaurant. Because it does not have a bar now the locals refer to it as 'halfa pub'. To get in, you drive around the lane at the back, through three metre high corrugated iron gates, and look for someone to serve you. The front is very imposing with steel bars and locked doors. At night the back gate is locked but if you're staying there and want to go out for a while, the owner lends you a key to get back in.

Dave, Carol, her parents Rob and Pat, and I arranged with the proprietors to have dinner later. Most then headed off for a hot shower but the wait was interminable. It takes 15 minutes to run the water until it gets hot. In the meantime you freeze in temperatures of about five degrees.

Chapter Twelve

Dave must have known something. He skipped the shower and went straight to the real pub. Carol and I went to meet him before dinner. There too, you go in the back. The locals congregate on the concrete and are served through a hole in the wall. The front bar does not have access to the back. Being a Friday there were a few people settling in for the night. Moses was perched on a stool and Dave had made friends with a few of the younger blokes.

They reckoned that the Culgoa had been flowing for a couple of weeks and we should have come down that way. We knew it was bone dry and that they were wrong. Some of them also reckoned that the Aboriginal fish traps had been built by the Chinese in the 1930s. It was all interesting conversation but not a lot of useful information came our way. If anything it simply confirmed the powerful way that myths take the place of actual observation.

Carol and I went back to the halfapub restaurant for dinner. There was wine available and the feed was good. Dave arrived right on time, had dinner and then went out to chase whatever action he could find. This was apparently around at the local RSL club where he had bought the Chinese food the previous night. Eight o'clock is early, but that was my bed time so while Dave rocked, I snored.

The following morning we all went out to launch the kayak. Carol and her parents wondered what the strange contraption was that looked like a tin on a stick. That's what it was – a tin on a stick. I had popped it there the previous afternoon to show Dave where to turn off the track. We took some group photos and they saw what this part of the journey was like. Seeing it in person was great for Carol. It had been less than a week since we had seen each other but descriptions and photos are not the same as actually being there.

The banks were about 15 metres high and sloped at 45 degrees. Because the water level had been higher and was now dropping, the edge was very muddy. "I don't like that, Dave," I said. "But here's the plan. You hold the back of the kayak while I hop in. The back wheels stay down and the front one is folded up. When I say so, give me a shove and I will launch across the mud and into the water."

It was a brilliant scheme. I couldn't understand why everyone protested that I was crazy! Actually, Carol didn't say that, she has had more than three decades to get used to me so she just shrugs her shoulders. Rolling down a 15 metre high bank is a bit scary but nothing like riding a big bike through bulldust at 100 kilometres an hour. More importantly, it worked: the kayak flew straight across the mud and out into the water.

Carol and her parents went to Bourke for the day. I paddled and negotiated log jams. Dave fished but caught nothing and did his usual stint of running. At the end of the day he collected me and we went back to the Middle in Brewarrina for another night in a bed and a Chinese feed at the RSL club. After Bourke, we were going to be too far away for Carol to visit. Maybe I could get back in eight weeks when we did a crew change. Eight weeks wouldn't be the end of the world, but we hadn't been apart for that long since before we were married on Australia Day in 1974. She and her parents left the next day at 7.30am and Dave drove me back down the river for another day of paddling.

The people at Brewarrina had been extremely friendly and it had been great to see family again, but we were focused on what lay ahead. This was the start of the isolated bit, where the support vehicle might not be able to get to the kayak. These remote areas also posed a problem of communication. The Telstra towers didn't cover all of the ground so there were some gaps. The support vehicle had to be within five kilometres of the kayak for the UHF to work. Luckily we had Farnie's satellite phone so, as long as the high gain Next G aerial on the vehicle was close enough to a tower, I could always call Dave in the ute from the satellite phone in the kayak. We also had the second satellite phone that Ken had organised, provided Dave turned it on, so that I could call satellite phone to satellite phone. We didn't have to resort to using this method anywhere on the trip however, as the coverage with the high gain aerial was pretty good.

Chapter Twelve

This morning, the river bank was even higher than the previous morning when we had proved that the kayak could roll down the bank and across the mud. With the air temperature at zero degrees it is unpleasant walking in mud in bare feet. It is also a bother washing muddy feet and putting shoes on in the kayak. Another fun type entry was called for.

Dave set up for filming. The kayak was right at the top of the bank. "What about that log in the water?" Dave asked. "Should miss it by 10 metres," I replied.

"Rolling," said Dave, to indicate that the camera was on. Away went the kayak, rocking and rolling and gathering speed. It wasn't going where it was aimed. It started to left turn towards the log. Leaning to the right was no help. The front hit the mud. The log was directly in front. It was a tree root curved into a loop. 'Fuck!' Straight into the loop went the nose. This held the whole front down while a surge of freezing cold, muddy water filled the cockpit. Wet balls, front of the kayak stuck under a loop of tree root, back stuck in the mud. It was all highly amusing to Dave. "Hole in one," he gleefully exclaimed. It took five minutes to extricate the kayak from its predicament.

After 10.00am the wind had dropped enough to light a fire. I pulled into the bank where there was some timber and some dry leaves and lit a fire. My clothes hung from sticks drying nicely while I munched away on the sandwiches which Carol had made back at the halfapub. It was a bit precarious doing all of this on a 45 degree slope but nothing fell into the water. The clothes dried rapidly so I dressed and set off again in less than half an hour.

There were often cattle and sheep beside the water, some wild pigs occasionally and ducks, swans and lots of other birds to keep me company. A white crane followed the kayak for two hours. At times the river was headed exactly in the wrong direction but then it would turn back. It took more than half an hour to travel round one bend to get 100m away in a straight line. At 1:250,000 the map of the river was only very approximate so it was impossible to tell by looking at it where I was.

We had arranged a radio call at 1.00pm when Dave would try to get close to the position where we estimated I would be. This worked fine so after we talked, Dave drove to a spot down the river where he could see me coming and I could see the ute and the camp.

It was a long way up the bank to Dave's camp site but it was a relief to leave the kayak down near the water, carrying up only the lunch bag, water bottle and GPS so the batteries could be charged. The river level seemed to have stopped falling, which was a good thing – any lower and it would be back to walking. Dave slept in the ute as usual but I took a tarp and hooked it over a branch. It was folded back under itself on the ground so it stopped dew from forming on the sleeping bag and it stopped the self inflating mattress from getting dirty. I could appreciate the beauty of the night sky much better than if I had been in a tent. Funny, you work as hard as you can all your life to get the things you want and it turns out that you can have it all for less than the basic pension.

In the morning we arranged to meet at the junction of the Barwon and the Culgoa Rivers. This is also the point that the Barwon becomes the Darling. Again we scheduled a radio contact for 1.00pm. Dave had to drive to Bourke on the south side of the river and then back up the north side where he was to meet the mayor of Bourke, Wayne O'Malley. Wayne had contacted us via Vikki, saying that we would be welcome to use the fishing hut on his property.

There were some really big pumps on this section of the river. We had heard many times that there were lots of pipes big enough to crawl up. Some of these would have almost been big enough to walk up. They were monsters. Dave checked with some irrigators and pumping is strictly controlled by water levels over the weirs at Brewarrina and at Bourke. Sections of the river were very clear, which was surprising. To see the tip of the paddle in the water, when most of the time it had disappeared in the same way that a spoon disappears into milk, was most pleasant. The temperature of the water was about that of a cold beer, nice if you want to keep your refreshments cool but not tempting for a swim or even a wash.

Chapter Twelve

About 9.00am I ran off the edge of our last 1:250,000 map. The new one was 1:100,000 so it showed the river in enough detail to follow every bend. About midday I took a photo of what looked like a large gully coming in from the right. I should have known where I was but I really had no idea. The bends were all shown on the map but I couldn't correlate the map features with real life. At 1.00pm we tried our scheduled radio hook up but Dave was very faint. We agreed to make contact again at 2.00pm. At 2.00pm there was no answer from Dave. "Shit! How can that be?" I decided to give it another half and hour and tried again at 2.30pm. Still there was no answer. This didn't make sense so I climbed the bank and tried again. Still nothing. Finally after I scaled a tree we made contact. I gave Dave my GPS co-ordinates which he fed into his GPS. "You are 3.7 kilometres away," he said. "Downstream! I will come and get you."

To make his overland drive a little easier, he uncoupled the trailer. At first the track was well used, then lightly used, then there was no track at all and Dave was testing his 4WD skills. Eventually I saw him through the trees on the far bank. It was too high and too steep to get the kayak up, so we moved downstream to the inside of the next bend. That section on the river was fine for paddling but very hostile for driving. Dave had a couple of ravines to negotiate as well as dealing with trees and ground littered with large logs. He helped me drag the kayak up the bank where we left it hidden among some logs.

The trailer was at Wayne's fishing hut. This was at the junction with the Culgoa. I had taken a photo of the junction as I passed but didn't recognise it for what it was. There was no water in the Culgoa. It is a deep gully with water backed up from the Darling for about 100 metres. The banks are steep. Unless you pick your path carefully it would be easy to slide all the way down. Both the Darling and the Culgoa here are like railway cuttings. Down the bottom is one world: dirty water, ducks, swans, birds in the trees far above, angular tree roots and high, grey banks. This is my world. Up the top is Dave's world: indecipherable tracks, trees to locate the river, multitudes of eroded gullies to negotiate, a ribbon of dirty water far below, almost inaccessible.

I decided that the tarp hooked to a tree would be more salubrious accommodation than the hut. Dave settled on the hut, claiming that staying in beds in Brewarrina had made him soft. There was a fire place in the hut, so he pulled an old bed over next to it and made sure the fire was well prepared.

While Dave was cooking up a storm in the cast iron pot over the open flame on the ground outside the hut, Wayne pulled in to say hello. The beer we offered him was only air temperature but as he later accepted a second one it probably hit the right spot. The property is 67,000 acres and Wayne has been in the area all his life, as were his father and his grandfather. Three generations give continuity to his observations.

Wayne explained that the clear water I had been seeing is saltier water that flows into the river from underneath. The locals use this term for water that comes in through the river bed, as opposed to flow from above ground by the side. The water that I had seen disappearing 1000 kilometres ago, back on the Darling Downs, was one part of this interaction between surface and ground water. Here I was at the opposite end. Both provide excellent examples of the interconnectedness of surface water and ground water. One cannot escape the fact that average annual rainfall for the area is just over 400 millimetres but that evaporation is 2,500 millimetres. With such a deficit, underground water plays a major role in the system.

Wayne thinks that for a river system in the midst of a drought, and a very bad drought at that, the Darling is remarkably good. He observes that twice as many gum trees grace the banks as did 40 years ago. The town weirs have maintained water using releases from the headwater storages, and the full weirs have benefited the trees and the fish. According to Wayne, if it were not for the weir pools, native fish stocks would have been devastated. Friends who grew up in Bourke and swam in the river half a century ago tell me they despair at what the town and the river have become. Trying to piece together what has happened since white men arrived is very difficult. So much has changed. Wayne's perspective is from looking at the trees, my despairing friends recall the quality of the swimming. Again we have different experiences giving rise to different observations.

Like Robert Buchan, the mayor of St George, Wayne is careful to explain that the towns are tied inextricably to the land. A six year drought is devastating to most farmers and is equally devastating to many of the town businesses. Between the 2002 and 2006 censuses the population of Bourke Shire fell from 4,100 to 3,200. In 1980, locals say, every major car dealership was represented in Bourke. Now there are none.

Wayne believes that there are undoubtedly climate changes in some parts of the world but the current drought is not due to climate change. He feels that we need to watch closely and assess the risk but not jump to any conclusions. This sounds logical, well balanced and well thought out. The problem is that the evidence is that climate change is happening very fast and people need to look at what action may be required to live with rising temperatures even if they can't do much to prevent them.

There was to be a council meeting on Monday of the following week. Although Dave and I had planned to be well downstream of Bourke by then we accepted the invitation to talk to councillors after the meeting. We would spend the nights in Bourke until then to take advantage of the opportunity.

The following morning we had a minor problem. It took half an hour of driving around, through wash outs, around trees, across bare paddocks, until we found where we had left the kayak. After waving goodbye, Dave returned to pack up the camp only to find that the tow ball had fallen off the ute, so he couldn't hook the trailer on. Retracing his tracks he eventually found it. In Bourke two days later, we had a locking nut fitted to prevent further episodes of the great towball hunt.

It was a day without logs. This was good. More water would have been nice because there were many places where the kayak was dragging along the mud. After morning tea I just pushed on, hoping things would get better. They did. From the start of the weir pool, about 45 kilometres upstream from the Bourke weir, it was just straight paddling. About 3.30pm there was Dave with the stove out, coffee ready and bacon grilling. My plan had been to eat my peanut butter sandwiches but Dave's hot bacon sandwiches were so much more attractive. Succumbing to that temptation meant a half hour rest but it was worth it. At the end of the day I had covered 51 kilometres, the first day over the milestone of 50 kilometres.

There are always two sides to every story and this is Dave's report that day.

> *"Steve did very well with the paddle and managed to punch out 51 kilometres on his way to Bourke. Whilst he was very keen to make it to North Bourke and possibly paddle through the late evening, sanity prevailed. I had set up a coffee and bacon sandwich stop by the side of the river which he was reluctant to stop at. After getting out with jelly legs (legs are used almost as much as the arms when paddling) and after a quick bite he agreed to do just a short distance further."*

We were north of Bourke by thirty river kilometres so the North Bourke caravan park with its hot showers was looking really good. The site was grassy as there were no restrictions on sprinklers. It was the first time Dave and I had tackled the big tent but we got it up easily and made ourselves very comfortable. The tent was to stay there until the following Wednesday, a period of seven days. 'Right luxury,' as the Monty Python men said in their famous skit.

The week in Bourke was a busy time for us with its history to learn, some schools to talk to, the distance education school, the big wharf on the river, council interviews and a couple of contacts to follow up. We felt very spoilt with such a comfortable big tent, complete with 240 volt power to run the computers and lights. We unpacked the Apple computer so that Dave and I could work out how to use it to store video clips. It took me a while to believe it, but all the clips that Ken had taken had been deleted. All of the photos that Ken had taken had been deleted. There were 1602 thumb images of Ken surfing, his son's wedding and other personal images. The main images were gone but the small versions were still there. So Ken and Barb had kept Carol's car for six days while they were downloading and deleting. And, they still had the external hard drive. This was not good.

Chapter Twelve

I called Ken the following morning, Friday. Firstly I said that as he was no longer a part of the team he couldn't expect 50 percent of any profits from a film but that he could have *pro rata* profits based on the percentage of footage he had taken that we used. He agreed that this was quite fair. With Dave listening, I enquired about what had happened to the still images.

> "I deleted them," he said.
> "Well, I'd like copies for our records," I said.
> "I will send them to you," was his response. It was as simple as that. No denial, no argument.

That is where we left it and Dave was the witness. I simply couldn't understand what I had done to make them do this. They had certainly let their resentment ferment and fester, yet when confronted Ken just agreed to send the photos.

There was another issue with images. We had taken photos of some of the kids at the schools after getting signed consent forms saying that this would be OK. Geoff had major legal and moral concerns though, about posting these images on the website. Geoff and I agreed that, forms or no forms, we wouldn't publish photographs of children. This was disappointing for the kids. We didn't like it, but the legal issues around those images being misused are complex. It is a sad comment on the world we live in these days.

We headed into Bourke for two school visits and to catch up with the shire engineer. His name is Allan Murdock and I had met him a couple of times in the past. He's a friend of Grant and Jenny Cobbin. It is a small world, this water industry, and if you play around in the business for a few decades you're bound to come across most of the people in the industry at some stage. Allan told us that the previous January, 108 pipes to properties had been broken. The kids break the pipes at the water meter and they get a good shower of water. It is apparently an impressive spray. He argued that the police were too busy to stop it happening and that 'a good kick up the arse' might solve some of these problems but there was no one who could do that.

On a side street in Bourke there was a lone shop that didn't have its windows shuttered or boarded up. Perhaps not being in the main street they felt safer. Perhaps being a hardware shop they were not worried. Whatever the reason, what most of us would call a normal shop front stood out like the proverbial dog's balls. Sadly, before the week was out the window was broken.

I don't believe that the grim nature of the streetscape is a true reflection of the town, or its people. As is the case in all these western towns, the Bourke locals are helpful and friendly. Putting shutters on everything after dark though, gives the impression of a town under siege. The impression is false. I feel completely safe walking around these towns at any time of the day or night. Sure, there are often indigenous people hanging around, but what else can they do if they can't get a job? They are always very amicable and chatty when we say hello. Even in the most run down of towns, if you stop to have a chat, everyone is terrific to talk to. Despite this personal goodwill, somebody smashes and robs the shops if they are not shuttered. People lament the fact that the place has gone backwards over the years and simply hope it will turn around some day. Despite many attempts to find an answer, no one has found it yet. I couldn't help thinking that the complete lack of opportunities and the feeling of hopelessness are behind it all.

Dave was keen to go to the Enngonia races on the Saturday. He dropped me in the river downstream of Bourke on Saturday morning, arranging to meet Sunday at 4.00pm. I was to call him on the satellite phone to give him an approximate location. This all worked a treat. Dave wasn't too effusive about what a great time he had in Enngonia so I suspect the raging party that he had hoped for didn't eventuate. My Saturday night on the river was very comfortable after I found a sandy beach to camp on. Breakfast was a wee bit ordinary though. I had forgotten to pack muesli and although a packet of dried peas and corn was very warming, it just didn't excite me. It also didn't provide energy for very long.

About a day and a half upstream of Bourke on the Wednesday we arrived at Wayne's place, I had met a group of people who were camping and fishing. There were up to 22 people there at times and some had stayed for two weeks. Amongst them they caught three yellow bellies and two cod in all that time. Now I came across another couple.

"Great set up you've got," I yelled. "Been here long?"
"Two months."
"Done any good with the fish?"
"Yeah, got ten."
Ten! That's what they said.

Two months, sixty days, two people. That's one fish for each person every twelve days. Even at one fish between two people every six days it is not what I'd call exciting. They seemed happy enough though. The people who enjoy the Darling are few and far between but are a real pleasure to meet.

The NSW Fisheries Department estimates that native fish stocks are at around 10 percent of pre-European settlement levels. Not many people would judge that to be satisfactory but there are plenty of people who will argue that all is fine and that we have no idea what fish populations were. Denial is everywhere. Some people will throw a net in, and some will use live bait knowing that this is illegal. Luckily, most do the right thing. Some people travel out from the big cities to experience the bush and try their luck with a fish, but the drought has reduced these visits to almost nothing.

Wayne had invited us to present to the councillors after the Monday Council meeting. Dave picked me up from the river at midday. We were at Rose Isle, about 70 kilometres by road from Bourke. Now that Ken and Barb had gone we wouldn't normally do that sort of driving, but it was important to speak to the Council. By 2.00pm we were set up and keen to get started. Introductions were cordial and we were both pleased to be able to contribute. After the great presenters we had worked with at Ipswich and Toowoomba there was no doubt that the information we had was up to date and state of the art.

The presentation has a number of slides relating to the implications of global warming. A critical slide is the one that describes the effect on extreme temperatures of moving the average slightly. An average rise of two degrees in Longreach means that the number of days over 40 degrees will increase from 17 currently to 54 by 2030. Unfortunately, we had no hard data for Bourke but had heard that the average number of days over 40 degrees is currently 12. Unclear how many they could expect by 2030, I recommended that Council seek information from the CSIRO. I also explained the difficulty of predicting future rainfall because the link between rainfall and average temperature is indirect.

One of my early slides features the book *The Weather Makers* by Tim Flannery. I nominate Tim's book as one reason I started thinking deeply about global warming. It was a bad error to say that. Many of the council had met Tim Flannery when he travelled down the Darling with his mate John Doyle. One of the councillors, Wally Mitchell, was a past mayor of Bourke and had spent time with Tim and John in Louth, 100 kilometres down the river. He was most unimpressed. Wally's attitude during questions was the most aggressive. "Don't come out here and tell us how to learn," he said. "We know how to learn. People are all excited about climate change but this is just a cycle." He quoted the years that the Darling had been dry before and pointed out that the current drought was similar to previous ones.

Wally, and the others at the meeting, knew precisely how much water flows down the Darling. They knew the potential yield from a dam on the Clarence River, near the coast across the Great Dividing Range. They believed that if only politicians at all levels would listen to them, the Clarence River could be diverted to make the inland prosperous.

We had stickers on the ute that said 'Not a Drop – Keep the Clarence Mighty'. Being born on the Clarence I know people who would die to save the river. No one in Bourke had noticed the stickers but my comment: "Over my dead body," pretty much settled it. I was a misguided greenie, just another one of the ignorant people who couldn't understand the value of diverting the Clarence River inland. Interestingly, they understood my determination not to transfer water from the Clarence to the Darling. They thought that, as a Grafton boy, it was only natural I would do everything in my power to hang onto 'my' water. My observation that all farmers are probably green at heart – they want their farms to be productive for generations – fell on deaf ears. The attitude of the Bourke Shire Council reflects a sign in the North Bourke pub, 'The only thing between a greenie's ears is wilderness.'

Luckily, Tuesday went better. Dave left for Brisbane, swapping vehicles with Jonathan at Moree, which is about half way. Jonathan was to do the next seven weeks as support crew. While the guys were away I was on the air at the Distance Education Centre having a great time with the kids.

This was the best school experience on the whole trip. The centre covers an area from the edge of the tablelands right out to Tilpa, 200 kilometres south west of Bourke. Nina, the teacher, talks into a camera to give the lesson. The students, at their properties, watch her on their computer screens. If they want to talk, they notify her via their keyboard, and their name is highlighted on the list on Nina's computer. That's how she knows who to ask to speak. Having brought up four kids of her own I dare say Nina knows a bit about kids. No doubt the challenges are different when most of the contact is by electronic media. It was exciting for me just being there and talking to them all. They were allowed one question each and it was great spending that time with them. I was to meet some of these kids later as we passed their properties down the river.

With Dave gone and some time on my hands at the tent there was time for some reflection on cotton. People say that cotton and rice use too much water for a dry continent. I'm inclined to agree with them but it isn't quite as simple as that. Cotton is am opportunistic crop to be sure. If there is no water, then don't plant it. If there is enough then away you go. Here at Bourke there had been no cotton planted for six years. But what about the citrus trees? They need water continually. Maybe they don't need as much but you can't stop watering them. The farms here were in big trouble with trees cut way back just to try to survive. Even when water comes there will not be a crop for a couple of years. Sometimes you just have to decide that citrus trees, grape vines, and water hungry crops like cotton simply do not belong in low rainfall areas.

The two weeks with Dave had been everything I had hoped. He had come along with less than 48 hours' notice and had experienced the Bokhara floodplain, Brewarrina, Bourke and downstream almost to Louth, including all the wonderful characters associated with these places. He revelled in the experience. One of the things that had stressed Ken, putting the movies onto a DVD, Dave achieved easily. The DVDs he made would play on any machine. This eliminated the previous problems of only being compatible with an Apple computer. The biggest change for me though was that Dave's primary goal was to support me to get as far as I could in the time that he was there. This was entirely different from what I had experienced with Ken and Barb. Dave and I were only a small team, but we were a team and part of a larger team. The trip was fun; I was relaxed and eagerly anticipating Jonathan's arrival. He arrived just on dark. We went to the North Bourke pub for the last time and talked about what might lie ahead.

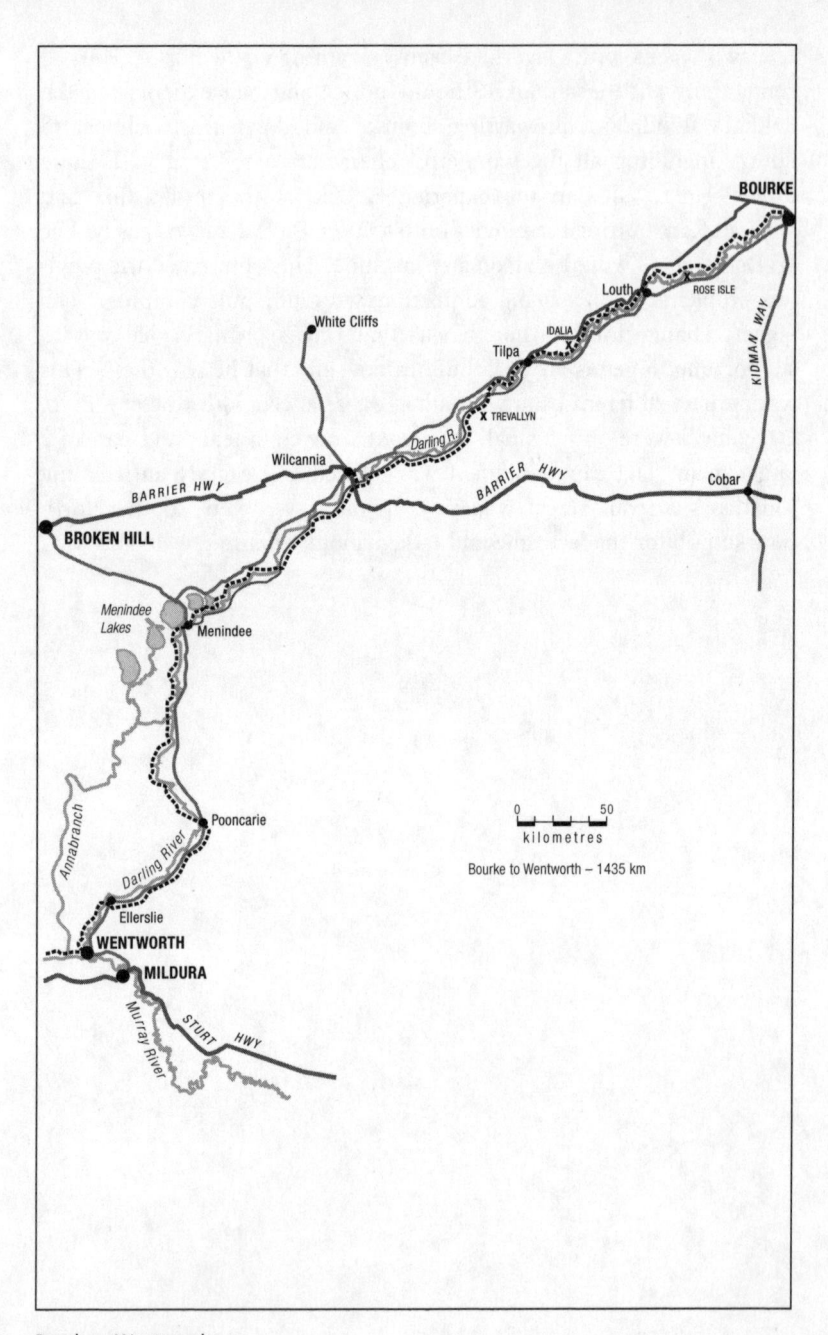

Bourke – Wentworth

180

CHAPTER THIRTEEN

Bourke – Trevallyn

DRIVING OUT TO GET STARTED BACK ON THE RIVER was Jonathan's first experience of outback dirt roads. It is bitumen all the way from Brisbane to Bourke but from there to Wilcannia, some 340 kilometres to the south west, it is all dirt. There are two types of dirt roads. The ones that have had gravel placed on them are fit for all weather use. Where there is no gravel, a 4WD vehicle can slip and slither along after up to 25 millimetres of rain but with more rain the road is practically impassable. Some of the dirt roads have signs that prohibit people from driving on them. Presumably if you drive on one of these roads you're breaking the law. It made me wonder what one does if it starts to rain when you are on the road.

We had stocked up at the supermarket, visited the Distance Education Centre and looked around town so it was late by the time we arrived at Rose Isle where Dave had collected me two days previously before the meeting with Bourke Council. Coincidentally, it was at a property owned by the daughter of our grumpy ex-mayor, Wally Mitchell, and her husband. She was a typical friendly country woman who had a wonderful garden around the house. It wasn't lush but it was a great retreat from the raw, parched country around the homestead. While expressing doubts about whether the garden was even justified, given the precious nature of water, she rationalized that some sort of escape from the harshness of the barren land preserved her sanity, allowing her to cope. It's hard to disagree with her.

Like father, like daughter: she was very firmly in the climate change denialist camp and expressed the opinion that people like her would be proved right within 20 years when the current cycle is over. I was learning not to argue. There is little point when the evidence is 20 years away.

We were lucky to find her at home. Many of the gates in the area are locked so her advice was invaluable in finding a way from one point to another. To get from her place to the weir where we had planned to camp the night, Jonathan had to drive the 40 kilometres to Louth, cross the river and come back up the north side road, then find his way through various property tracks. Without her advice this could not have happened. My journey for the day was less troublesome. Paddling through deep waterholes followed by the weir pool was easy. There was no dragging along the bottom for the whole day.

The river had changed since Bourke. It was wider, there were no snags – ie dead trees and logs – and the banks were free of any debris. There was plenty of dead timber at the top of the bank but it had all been cleaned out from the sides. All this was done many decades ago to allow easier passage of the river boats. It is sobering to realise just how long a simple action like that lasts in this environment. Nature acts very slowly here and cannot correct the folly of man even in a couple of lifetimes.

Occasionally there was evidence of punts that had been used to get across the river. The remains of rusting hulks and excavated bank entries tell of livelier times, of more people, more activity.

The river was increasingly clear. The murky grey waters typical of the Darling upstream more often than not gave way to clear sections where you could see to the tip of the paddle blade in the water. As Wayne O'Malley had said though, this clear water is salty water that flows in from the river bed. Mostly the salt is below the limits that would affect cattle. In fact, the green grass at some of the homesteads attests to the fact that the salinity is within the limits for most irrigation.

Upstream of the weir was a sign advising the intrepid traveller of the weir ahead. In Queensland the weirs were built such that you could drop two or more metres onto concrete posts designed to dissipate water energy but there were no signs to warn a kayaker coming downstream. Those Queensland weirs are not even on the maps. You just have to keep your wits about you, looking and listening for very subtle signs that there is a life threatening drop ahead. The weirs here in New South Wales were not life threatening. They were roughly the same height but the downstream side was packed with rocks arranged in a gentle slope away from the weir crest. Despite this they were well signposted. It's interesting how every bureaucracy develops its own culture.

Jonathan was at the weir when I arrived. He had planned on doing a 10 kilometre run, just like Dave had done, but did not have time. On the river I had no idea of the issues confronted by the support vehicle in getting to the designated point at the end of the day until we discussed it later. Over the years Jonathan and I had come to know and respect each other's abilities. We had ridden trail bikes together, raced each other and surfed with each other. I was, and am, more confident in his abilities than he is. He's very dependable and I was extremely pleased that he was my support crew through the isolated areas where toughness, both physical and mental, would be required.

This far into the Australian outback, the location of the Telstra towers is very important. Sometimes they are near a town but between the major centres they are spread more evenly without much regard for the small places. In the case of Louth or Tilpa, with populations that you can count on two hands, the towers are nowhere near the towns. Emails to update the website and calls to Carol and the support crew in Brisbane were therefore sporadic and unpredictable.

The task of filming and editing was picked up enthusiastically by Jonathan. Like Dave, he also immediately grasped the use of the digital SLR camera. The difference between us older folks and the younger ones who have grown up with digital technology is significant. We tend to take a shot of something. They take lots of shots. Professionals, in the main, take lots of shots and then pick the really good ones. In the time that Ken was taking the SLR photos he took about ten per week. With the younger guys operating the camera we finished up with more than 3000 photos for the trip. It is difficult to change the practices of a lifetime but given that you can delete the photos you don't want, it seems to me that the new way has merit. In this case, more is better.

Ken had posted the external hard drive to Carol a couple of weeks after he had returned the car and computer. Jonathan had brought it with him so once again we had computer space for editing. Ken had not, however, included the still images that he had deleted. Carol confirmed that she still had not received the images. This was starting to bother me.

The sunset at the weir was not as spectacular as some we had seen but Jonathan was enthusiastic so we have plenty of photos that capture the mood of the camp site. He seemed to be enjoying himself and adapting to the camp rituals. The fire needs to be close enough to the awning to provide warmth but not risk burning it. The water container needs to be away from the walking area so that if some water is spilt the camp does not become muddy. The computers need to be set up so that emails can be answered and film editing done. And then there is the shovel and the associated roll of toilet paper. You don't want to have to go searching for the shovel when the time comes.

To get down the weir, I had taken my shoes off and pushed the kayak through the rocks and water. The water was cold, colder than the beer in the trailer, so it was quite difficult and painful to walk on the slippery rocks under the water. This was the last time I tried that. For all the weirs downstream I kept my shoes on and my feet dry.

It was only 41 kilometres into Louth so the next day I arrived just after lunch. This was despite many shallow sections between the water holes. When there is no flow these shallow sections eventually dry up. It is possible to paddle if there is 100 millimetres of water. When it drops below that it is necessary to get out and slog through the mud or to stay in the kayak and push through by lifting your weight off the bottom. I found I could do this by resting the paddle on the kayak and using my hands on the bottom or by paddling accompanied with jerking movements of my bum to slide the kayak forwards. The bottom wriggling method was the one I usually adopted.

This is very hard work, so you gradually learn to read the river. Usually the shallowest spots are in the middle. You can't see them until you run aground. When the kayak stops there is a wave of water that briefly re-floats it but that usually just washes the kayak further into the shallows. Down one edge is the best place to go but picking which side will last the distance is tricky. To get it right more than 50 percent of the time indicates that it is not pure guesswork, there are signs that can be read. I got it right about three quarters of the time but I'm not exactly sure of what all the signs were. It felt like I was following my instinct and I'd love to know what signs were guiding me.

Jonathan had made arrangements to visit the Louth school the following morning so we set up camp on a neat grassy section of what aspires to be the caravan park. Louth gets very few visitors so the amenities are small demountables but they are comfortable and we didn't have to share the place with anyone. At sunset we went over to the golf course to enjoy Louth's unique and solitary tourist experience. There is a point on the golf course in line with a large cross on a headstone in the cemetery and a homestead down the river. Just as the sun touches the horizon, it shines on the cross so that it glows brightly like a huge light, reflecting the sunset. The position of the reflection moves slightly every day but on the birthday of the woman who is buried there, the light shines on the doorstep of the homestead she occupied.

In the pub there are always plenty of friendly locals to talk to as well as any passing trade, usually comprised of a few grey nomads camped on the other side of the river. The general store is located in the pub and occupies shelving about three metres wide by about two metres high. That is it, so it is not a place to stock up on supplies.

As usual, the school visit was the highlight of our time in the town and the kids loved the wombat presentation. The teachers are well aware of climate change and teach the principles of sustainability so the new generation understood the issues straight up. One of the teachers spoke about the area she was from. She described the importance of claypans on their property as being the sole source of water for stock dams. She also said that the gymkhana had recently been changed from September to July because the chance of getting hot weather in September was becoming too great. Despite this striking evidence of the reality of global warming, she cautioned me against being judgemental about farmers. "People out here don't believe in climate change, but you can't label them as ignorant. They just think about things differently," she said. Because she was intelligent, understood the issues and was teaching children the right information I took this on board.

On the matter of the claypans, though, I was still searching for the answer. The explanations provided by most people were at odds with guys like Greg Hoadley way back in Queensland and Graham at Bokhara. Just because three generations on the same property have the same idea does not necessarily make it right.

In the pub we had the same old discussion about river kilometres versus road kilometres. As usual, folklore ruled. The ratio is three to one. Everyone knows that. What we had recorded told a different story. The Brewarrina weir to Bourke weir is 216 kilometres as measured by the GPS on the kayak. This is two kilometres different from what the NSW Fisheries Department says. By road it is 105 kilometres. Those numbers indicate a river-to-road ratio of about 2:1. A sign near the river at Bourke says it is 1435 kilometres by river to Wentworth; road maps indicate 735km, again about 2:1. This covers the whole of the length of the Darling River and yet this message was discounted by the locals. As far as the pub was concerned, we just had to be wrong.

Of course these matters are immaterial, even petty, but they do illustrate the incredible power of folklore. As Graham says, people say that there were never any trees in some areas and yet when he takes the cattle off, trees appear. Farmers are very conservative. Their instinct is to resist change. The world is changing. It is changing rapidly and will need rapid responses. Climate change and peak oil will lead to slow changes over the longer term and dramatic fluctuations in the short term. The effect of wild fluctuations in oil prices will affect farmers who rely on diesel and oil derivatives such as fertilisers and pesticides. Both oil prices and weather directly influence food prices. It is going to take some very special people to lead city folk through the tumultuous times ahead, let alone country folk steeped in folklore and all the inertia that entails.

All of the school came down to the bridge to wave goodbye. Jonathan took photos and it was disappointing that we couldn't put them on the web site because of the fear of some creep using them maliciously. Heading away from Louth was indeed heading deeper into outback country. The next bitumen road we would see was 240 kilometres away. If the water in the river ran out, and we expected that to happen at some stage, it would be hard work pulling the kayak through the bulldust. For now, there was a very pleasant weir pool to paddle on between Louth and the Louth weir.

Finally Jonathan was to get time for a run. It was only 10 kilometres but it was a start and he knew now that he could work running into his routine. On top of that he had crammed a full set of weights in so that he did weight training to keep himself busy. The bar fitted nicely into the trailer with the round weights just fitting in front of the back seat of the ute.

Downstream of the weir is a property called Idalia. Dermot lives there. He's one of the young guys I had spoken to at the Distance Education day in Bourke with Nina. Idalia homestead is set back from the river. Although it was only a short walk, Dermot had ridden his motor bike over to talk with Jonathan and to wait for me. His mum, Jane, had placed a red tea towel on the bank two days earlier to show me where to stop. Luckily Dermot was waiting on the bank because I didn't see the tea towel. He and Jonathan had become mates and swapped stories about their lives and the country. Every Tuesday afternoon Dermot goes to the Louth school and once per term he goes to the Bourke school where all the distance education kids and families mix for three days. He loves his motorbike and the property and enjoys the freedom of the distance education system. He has a dedicated study room with the communication equipment in it.

I had met two other students from the Distance Education day, Hugo and Ollie, further upstream at Jandra Station. Hugo and Ollie had a dedicated room in a separate building and they had a tutor as well as periodic visits from Nina. Looking at the way the homesteads are set up it is obvious that the parents take education very seriously indeed and will go to great lengths for their children. The alternative is to send the kids to the city where it is impossible to grow up feeling for the land in the same way that it is living with it every day.

Before the recent small flow in the river on which I arrived, Hugo and Ollie had walked across the sandy river bed. They still had a water hole with plenty of water for the house but the river had stopped flowing and things were pretty grim on the property. Jandra has been in the hands of the same family for three generations. It is one of the very few homesteads that overlook the river. Not imposingly grand, it nonetheless has a spacious, homely feel that oozes history and quality. During floods in the past there had been concerns that it could be washed away so it is now anchored to prevent that happening.

Down here at Idalia you can't see the river from the homestead but everyone is very aware of the state of the river. There is neat farmstay accommodation and camping on the bank – the bank being the flat land next to the steep slope, like a railway cutting, down to the water. With all the media focus on the drought, people stopped coming in 2007. Jane struggles to understand why. She says that it is a window of opportunity to see the river dry in places, to see the rocks, sand bars and stones that are normally not visible. There are no fewer fish, in fact those that are there are more concentrated so she thinks it is crazy that the fishing groups have stopped coming.

Jane loves the bush life. She says it can be hard but it is worth it and she wouldn't swap life on the land for anything. The property is 59,000 acres with most income coming from wool. They have a fancy looking twin engine plane which I dare say brings in some income from charter flights. With the flights, the wool and the farmstay, they seem to make a go of things.

David Bristow joined us at Idalia. His company was a major sponsor of the trip and, as David is somewhat of a workaholic, his wife had suggested it would be a good break for him to come out and see us. It was also very timely so that I could talk about the Ken and photographs issue with someone not involved emotionally. David came with gifts including a bottle of port from him and one from Jenny Cobbin. Port is always a dangerous drink for me as it slips down way too easily at the end of a night. With a 50 something kilometre paddle the next day there was plenty of incentive to exercise restraint and I'm pleased to say that sanity prevailed. Over two nights though, the bottle was beginning to look the worse for wear.

Because Ken hadn't sent the photos as promised I had taken some action. Firstly, I left Jenny Cobbin with the task of trying to get them back. Secondly, I had asked for legal advice – both directly from our solicitor back home and from David, who had access to barristers via his family. All indications were that as Kayak4earth had paid for all of the expenses to enable the photos to be taken, the company would have a claim at law. I was not denying him his right to use them. I simply wanted to use them too. For him to deny me that right was little short of theft. The images had been taken for a purpose. Ken had signed onto Kayak4earth to achieve that purpose. Now he was obstructing that purpose. Not being able to confront Ken or to fight in any way was tough. I like to fight if I'm wronged. David's advice was not to think about it. "Steve, you are better than that," was his comment.

I knew that was the sensible thing to do, but could not let it rest. At the beginning of the trip I'd have said that personal politics was one of the things least likely to come up on this journey. Here I was finding the biggest challenge was a battle taking place in my mind. At least the physical surroundings gave me some context. The river was bigger than all of this nonsense.

The following day we woke to a magic sunrise over the Darling. It was the best one of the trip and we were pleased to share it with David.

CHAPTER THIRTEEN

Jonathan cooked bacon and eggs. His cooking really was something to look forward to. The whole of that day would be spent paddling within the Idalia property. David drove into Tilpa where he managed to get reception for his phone and to do some work. While there, he booked and paid for a room for Jonathan and me. A real bed for a night would be serious luxury.

Idalia has two different fish trap areas. Perhaps it was because the river was so low that I could see them. Perhaps it was because I had seen them back near Brewarrina and knew what I was looking at. Perhaps kayaking gives you plenty of time to take in detail. Whatever the answer, they offer a unique historical, almost spiritual, experience. It is as if the rocks can speak. They are positioned strategically within the structure of the river and the thought processes of the constructors thousands of years ago are obvious. The connection was palpable.

To get to the agreed camp site David and Jonathan kept together. My job was easy as every bend was shown on the 1:100,000 scale maps. It would be pretty hard to get totally lost on a river but it is a great deal more difficult above the banks. They made it without any problems though and just after I arrived we were all rewarded with another beautiful sunset.

David sums up his two nights as follows:

> *Both men were in fine spirits, dangerously healthy and enjoying the country fully. Jonathan cannot wait to pit himself against the elements, and Steve I think, is hoping the elements aren't home to hear his pleas.*
>
> *The river is amazing, I've travelled a lot of Australia and the country here is really unique, the shelter of the river gums and the river wet and flowing through some very dry woodland, clay pans and open plains, not the total desert country I was expecting to see.*
>
> *There are some great people keeping watch on the river and an eye out for Steve…his reputation precedes him.*
>
> *Jonathan is running a tidy camp with food that, while it takes his imagination a little time to concoct, was always welcome and finished to the metal plate every time.*

I spent two glorious days down the Darling with Steve and Jonathan…I even scouted out the Tilpa Pub…not a bad establishment for a population of six…cold beer, friendly landlady (Debbie) and the counter meal as good as Jono's cooking.

Steve has almost as many new questions about the river as he has had answered along the track, and local opinion is divided over his message, but as Banjo P once said, "he remains undaunted and his courage fiery hot…", don't know who's listening, but as I retraced Steve's steps this week back up the track, he is well remembered and the yarns abound.

Anyway, thanks Steve and Jonathan for being great hosts during my impromptu visit. The sunrises and sunsets were great and the yarns and ideas on the river even better.

In every pub that you visit, over the whole length of the river system, there is evidence of the amazing Murray Cod. Sometimes it is old photos. Sometimes it is the head and jaw mounted on the wall. Always the fish are huge. These huge cod used to be plentiful but are not now. Imagine the thrill when I saw a cod feeding. It was about half a metre long. There was just a tiny amount of vegetation along the bank, almost imperceptible, and the cod was grazing on it. As the kayak approached it rolled slowly to one side and with a thrust of the tail was gone. It felt like an encounter with an old man of the river, something spiritual, but at the same time something to quicken the pulse with excitement. Certainly it was an experience burned into the cells of the brain, never to be forgotten.

The Tilpa weir was a difficult beast. Down one side is steel sheetpiling and beside that are the remains of a fish ladder. Having resolved not to get my feet cold and wet again I stood balanced on the sheetpiling and tried to direct the kayak down the ladder using a stern rope. The cross members were rusted steel channels about 100 millimetres deep and half a metre above the cascading water. It looked easy at first but the kayak wouldn't go straight. It got hung up on the side posts and then it sagged between the cross members so that I had to climb down to move it, risking the steel collapsing under me. After half an hour of stubborn persistence the kayak was floating on the water below the ladder but it took a further 20 minutes to negotiate it through the rocks and into a pool about 100 metres downstream where I could paddle freely.

The Tilpa pub was much better than their difficult weir. It is worth travelling to. Perhaps paddling down the Darling to reach it is a bit extreme but it is definitely an outback experience not to be missed. It should be part of the grey nomad trail. In fine weather the roads are passable with a caravan but it seems only the very adventurous will go that way. For Jonathan and me it was a welcome beer, bed and feed as well as a chance to wash some clothes.

The paddle into Tilpa had been difficult with lots of bottom scraping and the bottom wriggling that goes with it. Debbie, the publican, said that the water had only arrived three days before us. Leaving Tilpa was not much better. There were long, deep pools but between them were shallow sections a few hundred metres long where the kayak would get stuck on sand or rocks.

At 5.30pm Jonathan was waiting at the camp site. He had set up where the road was close to the river for the last time for 25 kilometres. After that the river wandered away from the road at a 45 degree angle to about six kilometres between the river and the road. Each section between the pools seemed to be getting progressively shallower so it would only be a matter of time before the river was dry between the pools and I was back on the road again. Paddling was preferable to walking but if the water ran out in another 10 kilometres it would be a long haul back through country where there was not even a farm track. We decided not to risk this overland hike for the sake of what might be 20 kilometres paddling or might only be five.

Dragging the kayak up the bank was hard work. Previously, I had only taken the kayak out of the river where the banks were steep when I was going to put it back again the next day. The rules of the journey allowed the support crew to help with that because it was not part of the overall journey from Brisbane to Adelaide. Today though, clambering up the muddy bank with a six metre kayak was another part of the imaginary continuous line to Adelaide, just like going over or round a weir.

Since just below the Queensland border, the countryside had tinges of green and to the untrained eye it looked like the drought had broken. Everyone told us that although they were pleased to see this growth, it was winter broadleaf which would burn off in summer. They were optimistic that it may herald a summer where grasses would grow but as yet no one was saying the drought had broken.

The road was just natural dirt. Only the sections near cattle grids had gravel on them. Sometimes the dirt was hard packed and smooth but at others it was loose, soft and rough. My shoes were getting flogged and my heel was still sore. It hadn't recovered from the long stretches of walking across the plains of Queensland. That night I resolved to investigate the problem and used the biggest needle I could find to dig through about five millimetres of thick skin to a large infected area well under the skin. The release was not pretty but the relief was tremendous. It could only get better from here.

My shoulder was complaining a lot but then again it had been complaining while I was paddling as well. One night I woke up shouting in pain. It woke Jonathan as well which was highly embarrassing. That's the trouble with pain. During the day you can compartmentalise it, but when you're asleep you lose that control and the pain can grab you unawares.

Less than a year before it had been suggested to me that I might have a stiff shoulder.

> "What's a stiff shoulder?" I asked.
> "One that doesn't move," answered David, the orthopaedic surgeon.
> "Well that is not going to happen," I responded.

Given David's prognosis, I counted myself lucky that it had healed as well as it had. I had reflected on the shoulder quite recently because of an incident at the North Bourke pub. Dave – the hash man known

Chapter Thirteen

as Helga, not the orthopaedic surgeon – had taken to playing darts with the barmaid when things were quiet for her and invited me along. To pass the time in Libya I had often practised darts and played well in tournaments so I was keen to demonstrate my skills. It did not go as expected. All three darts hit the wall about a metre below the dart board. The pain was about a number eight. I excused myself and sat down. Dave suggested I throw with my left hand but I just wanted to be left alone. Maybe this was as good as the shoulder would get. Maybe yelling in the middle of the night would be normal. Maybe Carol would have to get used to that when I got home.

As if the elements hadn't slept well either, the morning started badly. After only 50 metres we had another flat tyre. Then after about two kilometres the front wheel collapsed. It was just a bearing, but the two repairs added up to a delay of over an hour. Although we were very isolated and thoughts of the city were far away, I still had a strong urge to push on as hard as possible.

Just after lunch we met Murray, who owns Trevallyn, a property with 80 kilometres of river frontage. Murray said there was water over at the river and we should check it out. We left the kayak beside the road and drove to the river which was only a few hundred metres away. There was a slight sprinkle of rain but it was not heavy enough to have any effect on the dust.

We were shocked at the sight. The river was covered in thick green scum. Maybe it was toxic blue green algae and maybe it wasn't. Neither of us was keen to try Amanda's method, which is to put your arm in it and see if it gets itchy. There may have been a couple of good waterholes to paddle on but the rough road, with all of its problems, was by far the more attractive option.

Murray had been on Trevallyn for 67 years. His opinions of people back up the river amused us. "Only been here 20 years," he would say. "What would he know?"

The family had records for over 100 years. They showed the river had been dry for 2300 days between 1900 and 1950. Between 1950 and 2000 it had been dry for only 700 days, despite the huge increase in extractions. Murray said that this was because their records also showed that the years 1950-2000 were much wetter than the previous 50 years. Murray Darling Basin Commission (MDBC) information suggests the river here flows at 57 percent of what it would without the extractions.

Early explorers, such as Sturt, had encountered water too salty for horses in areas around Brewarrina. We had been told that some of the clear sections could have salt as high as 8,000EC which is at the upper limits for horses and cattle and almost four times the level that humans can drink (humans can handle 3,000 EC in emergencies). Murray has been using water from the river all his life and has never noticed it to taste salty. When it is milky he said it tastes and feels fresh. When it is clear it tastes and feels hard.

The issue of claypans had been bothering me for the past thousand kilometres or more. Meeting Murray finally provided an answer that makes sense of the various facts and opinions I had collected. Murray reckons that in the 1950s you could land a Tiger Moth plane just about anywhere you liked. Now he has to look around to pick a spot to land his ultra-light plane. That means the claypans are getting smaller. He says they were caused by overstocking in the 1890s followed by the Federation drought. With wetter years since 1950 they have been receding.

This fits with Greg's idea of using his cattle to build up biomass in the soil. It fits with Graham's bio-organic methods which have clay pans disappearing on his property. It does not fit with Tony's idea that they are essential to keep coolibahs alive and it does not fit with the idea of people who claim they are essential to provide water for stock dams. It does not mean that their observations are inaccurate, just that they have put the incorrect interpretation on what they see. The claypans do capture water which supports the surrounding vegetation, but that does not make them a naturally occurring feature of the landscape.

Chapter Thirteen

I respected Murray's opinion considerably. It had been a long way down the Darling to this point and he was the first person we had struck who explained the clay pan issue in a way that made sense. One main thing that Murray, Jonathan and I agreed on after a wide ranging discussion: the river system cannot be governed effectively at state level. Murray thinks it had been a mistake at the time of Federation not to set the basin up under federal control.

Jonathan had broken the shovel that day while excavating for his morning task. Our attempt at humour involved plenty of boy type jokes about big jobs. We found that amusing but Murray was more practical and suggested he walk around a bit and find some soft ground in future. It didn't seem important at the time, but we were without a shovel.

The rain started to get heavier. By four o'clock the diameter of the wheels on the kayak had doubled from the build up of mud and I was 30 millimetres taller. The mud-engorged back wheels didn't cause any problems but the front wheel is housed in the frame that constantly shaves off the mud when it builds up to about 30 millimetres. As well as dragging a kayak over sticky ground I was powering a pottery wheel without purpose. It became impossible to go even the kilometre to the camp site.

After a call on the UHF Jonathan arrived with the ute. He was still in two wheel drive but mud was flying all over the place. We dropped the tailgate down and I sat on it, holding the harness frame. We could communicate via UHF but it was difficult to hold the kayak, the UHF radio and hang on to the ute. After years of riding a wave ski I reckon that my bum is trained to hold onto something pretty well. Despite my precarious hold on the tailgate, Jonathan tore off down the road. The mud on the kayak tyres was flying four metres in the air but everything worked beautifully and I stayed put.

We pulled into the camp site just outside a national park. The campsite was on a claypan but there were dead tree branches in the scrub nearby that we could collect for a fire. There isn't any noticeable difference between Murray's land, Trevallyn, and the park. It doesn't seem right to declare a national park in a landscape that was created by man's greed over a hundred years ago but here it is. There is a permanent ranger stationed there and another who comes out periodically from Wilcannia.

The rain continued. I decided to cook on the hot plate over the open fire. Jonathan was slightly smarter and stayed under the awning. He had placed the tent and the ute so that we could be under cover at all times. I had set up the billy to catch water off the awning. There had been no rain shown on our forecast but Murray has a link to a forecast with greater detail as part of his activities as a pilot and he had said that there may be 10 millimetres coming.

I woke at 1.00am. The tent had a puddle of water in it. Our sleeping bags were wet. The rain was drumming on the roof and it was now plopping onto puddles on the ground.

"You awake," I asked.
"Yep."
"You reckon we orta bale out?"
"Your call."

We lay there for a few minutes while I thought. The 10 millimetres of rain that Murray had predicted had certainly come, along with more. We were on a clay pan. There was no let up in sight and the rain seemed to be getting heavier. The shovel was broken so if we had to dig, we could be in real trouble.

"Let's go," I said.
"Righto."

The decision was made. It was cold, and muddy. Packing up took over an hour but we were finally ready. We locked the hubs in to engage four wheel drive. Neither of us knew what the driving conditions would be like but we opted for second gear, low ratio. My job was to stand in front and guide Jonathan up onto the road, a distance of about 100 metres. Jonathan had to build enough speed to make it up the slight incline to the road. It happened easily. Only the top 20 millimetres of the clay was wet. Below that it was dry so the tyres just threw off the slippery mud and bit onto the dry clay below. Once onto the road I clambered in beside Jonathan after scraping the worst of the mud off my shoes onto the running board.

The road was not as easy as the claypan. There was nothing under the mud. It was all mud. To stay on the road was a constant battle as the ute went almost sideways one way and then sideways the other way. Jonathan tried third gear low ratio which is getting close to a normal first gear but it didn't work. The sliding was too dangerous and to back off meant the revs dropped too low.

Driving required intense concentration. Our abilities are probably about equal in that regard. I watched as Jonathan learned the precise throttle he needed, when to back off, when to charge, when to counter steer and when to let it go. He was learning quickly and we were both thinking all would be fine.

We slipped off the road. All six wheels just slid to the left and down we went into a shallow table drain. Jonathan managed to slip into low gear as we went down. Fortunately there was just enough greenery to get some bite on the wheels. Ploughing on, we were both anxious not to stop. Jonathan tried to climb back onto the road after 100 metres but to no avail so we kept in a straight line. After about 400 metres he tried again and was successful. We had been on a slight bend and the road was cambered. That was why we had slipped off and why we couldn't get back up until the road was straight again.

Phew! What a relief.

> A cattle grid loomed at the extremity of the lights. "What are you going to do," I said.
> "Fucked if I know."
> "Can't do much except keep going and hope for the best," I advised.

It wouldn't be pretty to smash into the fence posts at the grid but it wouldn't be life threatening either.

About 10 metres before the grid, the tyres hit gravel. Everything straightened up and we were through. So that's what the gravel is for. Clever! Then it was sideways again and back to intense concentration. After about an hour I wondered whether we should swap drivers. Then I thought better of it. My logic was 'If we swap drivers we are back to the start of the learning curve. Jonathan has done the hard yards and been getting better and better.'

Thirty kilometres from Wilcannia the rain started to ease and we eventually got up to top gear low ratio. The driving got progressively easier and we even made it to high ratio. We had made it out of a potentially sticky situation that could have led to days of delay while the soil dried out.

The service station at Wilcannia was closed. It was 4.30am which meant we had to wait for two and a half hours and try to get some sleep in the car until it opened. Luckily for me, I was on the passenger side so there was no steering wheel to bother me. Neither seat would recline due to luggage in the back seat. We were both soaking wet and cold. Sleep was sporadic for me and non-existent for Jonathan. At 7.00am, two cups of coffee and a plate of bacon and eggs each brought us back to life for a two hour drive to Broken Hill. There was not enough water at Wilcannia to get the mud off the ute and trailer.

The shower in the caravan park at Broken Hill was bliss. We felt human again. There is nothing like a western shower. The water flows out in torrents. Water saving shower heads are like hens' teeth. I have noticed over many years that the best showers are west of the Great Divide. The mud, the dirt, the sweat, all sneak up on you gradually and you come to accept it. A hot shower reminds you of what civilisation means.

We erected the large tent, water-blasted the mud off the vehicle, fixed the inverter so that we could have 240 volt power out of the ute again, did the washing and ate at the pub. It was luxury for two days while we waited for the ground to dry out. With a strong Telstra signal we both caught up on emails and updated the web site information for Geoff.

Chapter Thirteen

Broken Hill was a surprising cultural shock. Neither of us expected to be impressed with the city. Bitumen streets, shops, supermarkets, garages, clubs, pubs, fast food, all the things that make up a city were there. This was a far cry from a wet clay pan next to a mud road. At first we were excited and soaked it all up enthusiastically but in the end were glad to depart civilisation and get back to the bush.

Arriving in town on Friday had been very handy as the trades businesses were all open. By the time we left at Sunday lunch time we had managed to repair everything that had been broken and had even cut down the table. The table was made to clip onto the side of the trailer but had been slightly too long to store in the back of the ute. By cutting it down it could slide into the ute where it took up negligible space and wouldn't be the nuisance it had been. And the most important thing: we had a new shovel.

Chapter Fourteen

Trevallyn – Wentworth

As we drove back up the dirt to Trevallyn, we congratulated ourselves on our timing. There were just a few wet patches which would be totally dry by the morning. Our escape tracks were obvious on the now hard clay but they weren't nearly as deep as some that we had seen in other areas. Jonathan said that after an hour of driving in the mud, he had been concerned that his concentration was slipping but had decided not to ask me to take over. It is interesting that we had both thought the same thing at the same time. It had definitely been a blessing that Ken and Barb had quit and Jonathan was with me.

The next day was tough. It was cold and windy but the trouble was the balls of my feet. Jonathan seemed fine, running his usual 15 kilometres and then doing his weight training. My new size 10 shoes were falling to pieces. Luckily Jonathan had picked a great camp site where there was some sand mixed in with the clay, respectable trees and even some grass. After two beers the pain had gone and the world was fine again. We had covered 33 kilometres, only another 30 kilometres to Wilcannia.

The following day was equally tough. Maybe when I thought my feet had recovered during the paddle from Brewarrina, I had been deluding myself. Maybe the rough roads and the load of the kayak created the problem. Whatever it was, the bitumen was an extremely welcome sight when it appeared mid-afternoon. A truckie called Duck pulled up and gave me a can of bourbon mixed at eight percent alcohol to drink at the end of the day. When I drank it two days later I had to agree with him. It was one of the nicest drinks I have ever had.

On the bridge at Wilcannia we met a teacher from the local Christian school. We arranged to do a presentation to her kids the following day. Jonathan had already booked us at the state school in the morning and there was a Catchment Management Committee meeting in the afternoon that I wanted to attend. That filled up the day.

The camping area was away from the town and there were no other campers there. It would have been fine except that we would need to leave our gear there unattended a few times. The motel was behind the service station which had a locked yard with dogs where we could leave the trailer and kayak. Two nights in a motel was extravagant so closely following the two nights at Broken Hill caravan park but less costly than losing critical gear. We couldn't risk leaving anything unattended in the open.

There was no water in the river but the Christian school wanted to do the presentation on the river bank anyway. It worked fine but it did seem a little incongruous with just sand and some bits of green grass for a river.

Wilcannia has been a thriving town in the past. Even as late as the 1970s the locals say that it was a busy little place. Now it looks like it is dead but it won't lie down. There are not enough people left to run the place. An 80 year old woman does the books for two organisations because there is no one else.

The buildings in Wilcannia are barricaded with bars and galvanized iron sheeting just like the ones we had seen in towns across this corner of New South Wales. Caravaners shun the town. Grey nomads tell each other not to stop there. The town looks scary. Again, like all of the western New South Wales towns we had seen, looks can be deceptive. The Catchment Management Authority meeting was held in a large room in the Westpac bank. After it was over I needed to return to ask for directions. I walked in, around the counter and up the hall. If I had tried that in a bank in a city I'd probably have been shot. It is a very friendly, relaxed town but the iron bars and shutters certainly do not give that impression.

A comment from one of the local women rang true with us. She said that when the river is flowing the kids are always there, swimming, jumping off the bridge, swinging from trees and generally dissipating a lot of energy. Without the river they are unhappy and cause problems.

Interviewing a Department of Fisheries scientist, we asked what changes she would like to see in her lifetime. There are four species of fish on the endangered list. Her ambition is to get them off that list. Although optimistic, she's a realist and concedes that the task is enormous. It is a pretty sad state of affairs when the odds are stacked against achieving even such a moderate goal.

An ABC journalist from Broken Hill had been in touch, wanting to do an interview about the trip. We met on the road outside Wilcannia and recorded quite a long interview. At the end, when he asked what was next, I said that it was too early to say but there was a bigger and better trip that beckoned. At the time it seemed like a pretty harmless statement but Carol heard the interview. "So, what's next," she asked when I spoke to her on the phone. "Are you going to let me in on this?"

Carol's request was quite reasonable, I suppose, but these sorts of questions can be uncomfortable. It is a bit like explaining that every man needs two motor bikes, or three sea kayaks plus a racing kayak plus a double wave ski and a single wave ski. And then there is the spare wave ski but let's not go there. What seems quite logical to men is often not apparent to their wives and further justification only seems to make matters worse. Sometimes it is better not to try. 'When you find yourself in a hole, stop digging.'

The people of Wilcannia thought the same way about the river as many of the farmers we had spoken to. It had been dry before, many times, but the normal mid to high flows didn't seem to be there any more.

Chapter Fourteen

On top of all the other woes, a thing called woody weed seems to be invading the area. Woody weed is a bush. There are many different varieties but the definition common to them all is that they have 'hard lignified tissues especially stems'. They are prickly, contribute to erosion, take up areas where pasture should be and thus significantly reduce stocking capacity. As yet, no use has been found for them. There is some research being carried out into using the woody weed as a feedstock for biofuel production but this seems a long way off. There are a number of barriers to biofuel in general, some due to the technology, some the application of that technology. In most cases it is not the technology that holds us back, it is the management of that technology. To think that technology is going to solve world fuel, food and water shortages is naïve folly in the extreme.

The Darling was starting to become a grind. It was a long way. Thoughts of the Southern Ocean had faded and we started to fantasise about Wentworth and the Murray River. That was still 386 kilometres away by road, twice that by river. Marching south west down a rough dirt road was at least going in the right direction.

The road south of Wilcannia is grey. The land is grey. The woody weed is grey. Even the trees seem grey. My mate Rod, fellow engineer but part-time artist, complained that none of our photos had any colour. I was not about to try and explain to him this is what happens if there is no colour available, he and I have shared too many bottles of wine. If he wanted colour then we would provide it. That evening we stoked the fire up before dark and took some shots. 'There you are Rod, here is some colour for you,' I thought. The background might be grey but the fire was yellow and orange.

Grey roads and no rain mean bulldust. Bulldust is fine particles of clay. It is a bit like sand but much finer, more like talcum powder. It flows around your feet and gets into everything. A major concern at the outset was whether the kayak could be pulled through bulldust. The answer is yes, it can, but it is *very* hard work. To look in front of you and see another 300 metres of even 50 millimetres thick bulldust is heartbreaking. The side of the road was better at times but the uneven ground there presented a different set of problems. My endurance was being taxed and that made other aspects of the journey, such as filming, difficult. Dave had learned this when he was with me and now Jonathan was starting to understand. It is much better to film when I'm not so tired.

Rod must have influence with someone, somewhere, because the following day we came across red sand hills. This was the spectacular outback colour that he wanted. If you have ever walked up a sand hill you may begin to understand what it means to drag 70 kilograms of kayak up one. 'Bugger Rod and his colour. I could do without these things,' I thought.

The locals had all told us that there should be water at Tintinalogy. This was 75 kilometres from Wilcannia. If there was water, I had only three days of paddling after that and we would be in Menindee. The GPS showed that it was only 47 kilometres in a straight line – pity about the bends. Luckily the locals were right. Tintinalogy was the start of the tailwater from the dam at Menindee.

The 'Tintin' camp site was a kilometre from the road but right on the river bank. Although camped only 30 steps from the water it took me 45 minutes the next morning to get there. I got halfway to the river and then a tyre went flat. The problem was exacerbated by a collapsed bearing which had to be replaced as well. Two rugged days in a row may have contributed to the gear failure. Since I don't need the wheels on the water, it was tempting to leave repairs until later. I took the precaution of fixing it properly because you never know what might lie ahead.

The next camp site was 46 kilometres down the river. Jonathan arrived early, set it up and then went for his run. He was a bit concerned about a huge goanna in the campsite but it ambled away after eyeing him up and down a few times. The sleeping arrangements for the night were Jonathan in the ute and me under the awning. That made the goanna my problem. I figured it would mind its own business and didn't give it a second thought. On his run, Jonathan saw a snake sunning itself on the road. The weather was starting to warm up and the reptiles were moving. They were slow but that didn't stop Jonathan leaping about two metres in the air when he saw it. I was happy for him to amuse himself with thoughts of a goanna sniffing around my sleeping bag, as I enjoyed my image of him running horizontally with an immediate transition to vertical.

There was water in the river backed up from the weir at Menindee, which is beside the Menindee Lakes. The map showed that this river backup joined Lake Wetherill. On setting out the following morning we arranged to meet at Windalle at the start of Lake Wetherill, which dominated our map. The path of the river is shown within the outline of the lake. Obviously this was where the river was before the dam was built. It was going to be fun to paddle in a straight line, down the edge of the lake.

Half way to Lake Wetherill I spotted what looked like a shortcut, which the map indicated could save about two kilometres. Until you get to these smaller features, you never know if they will be worth exploring. Sometimes a cut-through is three metres above the water level. Luckily this one was low down and full of water so my spirits soared. It would be an early rendezvous with Jonathan.

Just before the river reaches the lake, it heads north and then turns back on itself to enter the lake. At that point the map shows the lake and the river almost touching. The distance between them couldn't be more than 100 metres. Flushed with the previous success, I decided to cut the corner. That would leave me a quick paddle of about a kilometre to the rendezvous point.

After 50 metres crossing the short cut, the ground opened into huge cracks 150 millimetres wide and 300 millimetres deep. The kayak would bottom out as it crashed through these. 'Not to worry,' I thought. 'It can't be far.'

But it was. I was starting to get confused. Where there should have been water there was none. I called Jonathan on the radio.

    "What's the lake look like where you are?" I asked.
    "It's not very wide. Looks just like the river."
    "Shit. Then there's no lake."
    "Well no. Not here there isn't."

Pushing on I came to the embankment marked on the map. It looked like a levee bank to contain the water. Whatever water there was, it was a long way from the levee. On the other side the ground was sandy and a wide road appeared. I discovered that it was the main road from Wilcannia when Jonathan drove out to meet me.

"What are you doing here?" he said. This was probably a logical question but I replied, "Don't ask. How far along the road is it to the water?"

"You're nearly there. It's probably less than a kilometre."

Without the harness the sandy road was really difficult but not half as bad as the cracks and holes back out on the dry lake bed where I had expected to paddle. I trudged to the river and paddled a few more bends before calling it a day when the river came near the road again.

Jonathan had the tent erected at a camp ground on the edge of Lake Menindee. We were on a hill overlooking the lake. The scene is no doubt quite picturesque sometimes, watching the sun set across the lake with the opposite shore line barely visible. The scene was spoiled because there was no water – none. I was keen for a beer and a feed at the pub so we headed into town.

We ordered schnitzel which came with mountains of vegetables on a huge plate. With these plates demolished in a few minutes I asked Jonathan if he was satisfied. "I'm always hungry," he said. Despite valiant efforts to find the dessert menu there was none, so I ordered lasagne for us. Again it was a huge plate with chips and vegies. After a second huge meal I was replete and I think even Jonathan may have been satisfied.

For the past few weeks I had become used to being hungry. I reckon the only way to lose weight is to learn how to be hungry. It's not so bad once you get used to an empty stomach. There was no fat left on my body so I had no reserves to eat into and was consuming all the energy I could

Chapter Fourteen

get from my food. It had taken a long time to reach this point. For almost two thousand kilometres, my fat reserves had been a useful source of energy but it was pretty clear that stage of the trip was over. The two meat pies I had at the Hebel pub after dinner on July 1st had been the start of a significant change.

After a shower at the caravan park I collapsed into the sleeping bag and forgot to charge the UHF radio battery. "That's OK," said Jonathan. "Use the VHF." Despite my protestations, it worked fine, probably because the land was very flat. It was a river with trees on low banks that obviously were flooded from time to time when the area became a lake.

The lakes have always filled and drained. They are like a balancing system, regulating downstream flows. During floods they fill up and then drain out, making the flow into the Murray more regular. Since the construction of dams and control gates on them, the function of the lakes has changed. They are now used more as a water storage, especially for the town of Broken Hill. Significant horticulture has grown up around Menindee to make use of the irrigation water. It seems incongruous in an area that feels like a desert, to come across orchards or vineyards. I found it quite confronting.

On the river there were hundreds of birds. Birds had been a highlight of the trip. Just after Bourke over a hundred pelicans had taken off over the top of me. Fumbling with the camera in my excitement, I missed getting a decent shot. Even when I took my time later it was never possible to convey the sensation of all these water birds. Pelicans and shags are water birds, not seaside birds as I had thought. There were a lot more here than I had ever seen at the beach. Ducks always fly away from the kayak, swans usually do but not always. Shags sometimes fly away and sometimes fly back towards the kayak. In most cases the pelicans fly towards the kayak, grunting as they go. They are probably checking it out. When a hundred shags and a hundred pelicans take off and head towards you, it is always a worry. You can see the white trails of bird shit spraying across the water. Luckily they never seem to fly directly overhead so the hiss of their droppings on the water was always at least two metres away.

For two days I paddled the backed up water from the weir.. The river looked like a great place for houseboats but there were none to be seen. I think this is a potential business opportunity: outback houseboat hire. You would have plenty of solitude plus a truly beautiful river to play on. The boats would have to be specially designed for the narrow sections but I already had many designs in my head.

Although the area was blue on the map, indicating a lake, there were lines that helped me establish where the river went. A few kilometres from the dam wall there was a channel that was not shown. It cut off a huge loop. Not knowing for sure what it was, I cut through it anyway. Where it joined back to the main river a tourist boat, the River Lady, appeared. This was the first craft we had seen on the water and it was totally unexpected. It headed up river and we waved to each other.

About two kilometres further on it was time for a radio check. The VHF worked fine. Jonathan was waiting at the weir. The River Lady had turned and was coming back. 'Let's give him a run for his money,' I thought and set off as fast as I could go. They might overtake me but a race is a race and anything that is a challenge is good. We arrived at Jonathan together. The race was a great way to get me going and to pass the time quickly.

The owners of the River Lady, Pat and Malcolm, get between 2,000 and 3,000 tourists per year. Pat said she would put a sign out at their house 25 kilometres below the dam and I should stop for a cup of tea. We got their address as well so Jonathan was able to find it and meet me there just before dark. They are both passionate about the Darling and Malcolm is optimistic that we may have finally hit bottom in terms of stuffing it up.

Weir 32 is about 20 kilometres downstream of Menindee. Its pool backs up through the town and supplies the water that is withdrawn for Broken Hill. Permanent water changes the river characteristics in many ways. Pat and Malcolm have a pontoon, as do many of the residents. There is something about the permanence of the water that instils empathy, a love for the river. This empathy is very rare above Menindee except with the kids and a handful of the adults in Wilcannia and Bourke.

Chapter Fourteen

Malcolm wants the river to be natural again. He thinks that greed and politics are to blame for its destruction. A couple of years ago there was an inappropriate release of water from the dam and thousands of cod were killed. Some were almost 100 years old. Interestingly, he says that a dry river brings violence to some of the people, echoing the comments of the woman from Wilcannia. Malcolm lives on a weir pool: what would his reaction be if Weir 32 was demolished?

Menindee School was an unexpected surprise. The staff are very dedicated and the headmaster speaks a bit like a politician. They have a feisty attitude and involve the students in programs that support river health. It was perhaps this feistiness that was the most impressive. Whatever it was, the school received our data projector and screen at the end of the trip. There were other very worthy schools but most had the equipment anyway. Menindee got it on the basis that it will be used to promote the environment in school activities.

When we were in the pub we enquired about what sort of weather we could expect. So far it had been mainly cold interspersed with some hot periods. Jonathan had enjoyed the 25 degree days more than the 15 degree days but he was well aware that the warmer weather meant more reptiles. The winds had been strong most of the time and it was very rare for them to be favourable. They were almost always straight in the face. The answer was not good. 'Probably until about the end of September they will be mainly strong south westerlies,' was the local advice.

'Great,' I thought. We were headed south west!

Because we had the school visit and an appointment with the local water regulator we decided to keep the tent at the caravan park for another day. Jonathan located me downriver well after the sun had gone down. It was quite isolated and tracks were difficult for him to find. There were some pretty big roos about as well, so we worried that we might hit one.

With the visits, I had only covered 28.4 kilometres for the day but we made up for it the next day. Guided by Jonathan's torch I pulled in at

6.40pm. It had been a struggle not to crash into trees but the camping point had been pre-arranged. I just had to get there. It was 66.8 kilometres which smashed the previous best by 10.4 kilometres. The following day brought 35 knots of wind, mainly on the nose. It was tough going but at 48.2 kilometres for the day it was still very satisfying. The next day was 54.4 kilometres and finally 42.2 kilometres into Pooncarie. After weeks of grindingly slow work across country I was finally on the water and making good progress.

After day four from Menindee Jonathan asked me if we stank.

"Not that I've noticed," I said. "Apart from your bloody farts."
"That's what I thought, but I wasn't sure. Thought we might just be used to it."

I was changing my shirt after three days whether it needed to be changed or not. Of course I slept in it as well, but we did have sleeping bag liners. Jonathan went ahead on the last day and had a shower. The next morning he reminded me of our conversation.

His verdict: "You stink Dad."
"What of?"
"Dunno, everything maybe. You just stink".

Well that was pretty clear. The answer was to only hang around with other people who stank as well.

At Pooncarie the water was about 300 millimetres below the top of the weir but until I had reached the weir pool there had been some flow. That meant the weir was filling with the water being released from Lake Wetherill. This part of the river featured a new obstacle; blockages caused by reeds growing from bank to bank. It was an entirely different problem but it gave me something new to think about. If there was vegetation in the river here in parts, why was there none above the weir pool at Menindee Lakes and back up to Bourke?

According to some locals rainfall had been average for the past two years. They only expect 225 to 250 millimetres per year and that is what they had got. Geologists and other mining type people were staying in the pub as the area was being opened up to exploit some mineral deposits.

Typically, the kids at the school loved the kayak. They thought the wombat meme was great; we had been showing him since we first hit the outback in Surat and he received a good reception every time. Less typical was their depth of understanding about sustainability. When the river dried up a couple of years back, many of their parents had put in bores. Apparently these had become salty and had started to affect the waterholes in the river so the government closed them. As a result, these children understood first hand what the wombat means when he says 'everything is connected.'

With a late start after the school visit, we camped that evening at 30 kilometres. I wasn't worried because we had talked about completing the official marathon distance of 42.2 kilometres the following day. While I dragged the kayak through a marathon Jonathan would go into Wentworth to put the big tent up. Wentworth marked the end of the Darling and the beginning of the Murray. We were going to base ourselves there for three or four days as a kind of holiday and to take the opportunity to repair some equipment. Our web site blogs give the two perspectives on the day.

> *Steve:* *We set out for a big day today and managed 42.4km. I ran a marathon in 3hrs 52mins 15 years ago and the same distance today, towing a kayak took 9hrs 55mins. Recovery was a lot slower than usual. My feet really hurt.*
>
> *Talking about recovery, that is the secret to this sort of thing. At the end of the day I cannot walk properly. This has been the case from the start. But after a rest things start to work again. This time it took longer and I think things are a bit harder as I have lost all my fat reserves. I need to eat about twice as much at night as I did in the first couple of months. The amount that Jonathan and I eat is incredible. On the road, or the river, Jonathan piles my plate up with vegies, meat and gravy and I eat the lot every night.*

Jonathan: *With his intention of achieving his personal milestone of 42.2 kilometres (the distance of a marathon) for the day, Steve set off at 7:30am. He had a couple more refuelling stops than usual to keep the nutrition levels up and a few more chocolate bars than normal but managed to cross the 42.4 kilometres mark at 5:30 pm. Being his son and knowing his determined character I knew he was always going to make it. I think if his legs or feet did decide to give in he would have crawled to the 42.2 kilometres point. Thankfully, nothing is injured, however he did look like he needed a walking frame to order a middy from the bar later that evening.*

After that effort, the rest day on the Murray was very welcome and the city of Mildura fascinating after so long in the bush. After a sneak preview of the Murray, a quick visit to Victoria – our third state on this trip – and a restful day, it was back out to the road: walk to Ellerslie and then paddle down the last leg of the Darling to Wentworth. At the Ellerslie School we decided to be a bit daring for the kids so, after the usual presentation, we headed over to the river with the whole school in tow. This was the upper reaches of the water backed up from Lock 10 on the Murray so there was plenty of it and it was also beautifully clear. The bank was steep but we had successfully practised these dramatic, high speed entries a few times. Jonathan set up the camera and I set up the kayak. The front wheel was folded up and one of the back wheels was hooked on a tree. The banks were a bit steeper than the last time we tried this, so I hoped that there would be no stuff ups. Hurting myself in front of a whole school would not be good form.

> Climbing in I shouted to the kids, "Are you ready yet?"
> "Yes," came the chorus.
> "I'm not, I'm scared!" It was meant as a joke but was pretty close to the truth.

There was much laughter as I launched. Yelling "Whoah," I careened down the bank, into the water, submerging the front three metres of the kayak and then, thankfully, popping up.

Chapter Fourteen

"Phew. Pulled it off," I thought. Funny what one does for excitement sometimes. The school loved it, Jonathan was amused, and we had good footage of me being a dickhead. It was all good as I set off down river for Wentworth.

It is more than 1400 kilometres from Bourke to the Murray River. This is a long way down a river that is like paddling in a railway cutting. About halfway down this distance I had started to think about getting the Darling over and done with. It just seemed to go on and on. With a dry river, red sand hills, bulldust, algal blooms and isolation, all thoughts of the Southern Ocean had slipped out of my mind to be replaced with the goal of completing the Darling. Very few people paddle the Darling. Many paddle the Murray, and Tammy Van Wisse even swam the whole river. What would it be like to see the locks, permanent water, large towns and whole industries based on water from the river? It seemed akin to finding an oasis in the desert.

Part Three

The water from the mountains of the east coast has travelled down these rivers, through a lake and the Coorong then out to sea for millions of years. Then we got clever.

Wentworth – Adelaide

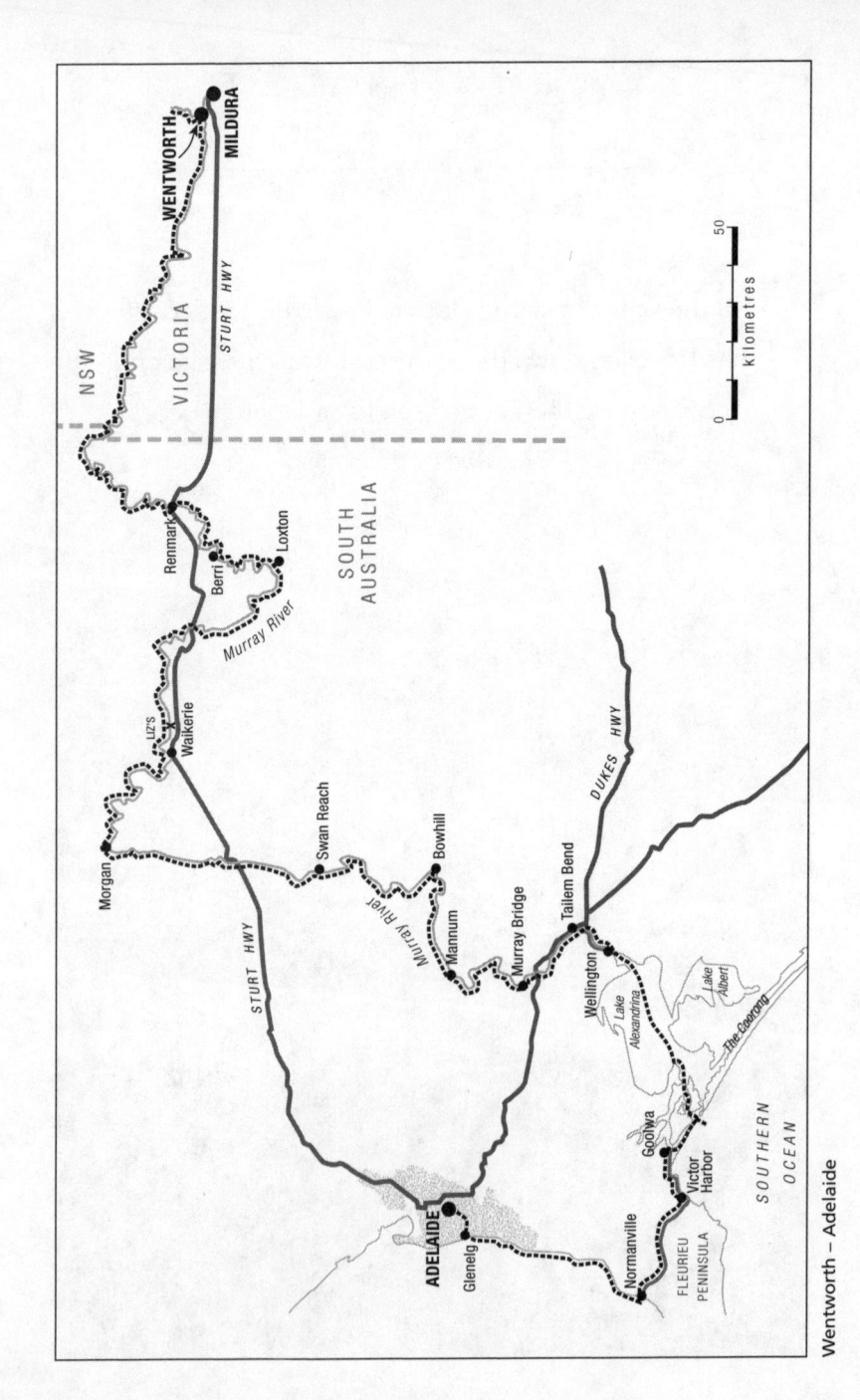

CHAPTER FIFTEEN

Onto the Murray

TO COME DOWN THE DARLING RIVER and arrive at the Murray River is a powerful experience. From a narrow river that feels like a railway cutting to a huge expanse of water with very low banks almost defies comprehension. The spirit soars. It was definitely a milestone. The accomplishment of getting there could never be taken away.

Although this was my personal perspective, the Murray has far greater significance to almost every Australian. All of the rivers in inland south eastern Australia flow to the Murray. For much of its length it forms the border between New South Wales and Victoria, Australia's most populous states. Not only is it Australia's longest river, it is probably its most iconic. To facilitate navigation it has 13 locks which means that it never runs dry like the Darling. People talk about it, travel on it, fish in it, write about it and generally just love it. Yes, it would be good to reach the Murray.

The camp site Jonathan had chosen was near the junction of the Darling and the Murray. The park was shaded and grassy. To get the kayak onto the bank, you simply paddled flat out and glided up onto the grass. The combination of easy access to the river, the lush grass and the luxury of the big tent was simply heaven. The caravan site seemed like a resort to us and it was cheap. The owners supported what we were doing and gave us a really good deal. With the water lapping at the front of the tent, ducks feeding in front of us and possums to play with at night, Jonathan had plenty to do.

Chapter Fifteen

Allen and Mary, two long time friends from the Australian Water Association, came from Shepparton for a night, and stayed next to us in a cabin that actually extends over the water. When Allen was president he was responsible for commercialising the Association to cope with the incoming Goods and Services Tax. He was also part of the team responsible for changing the name of the association to its current form, although Greg Cawston nailed it down as president at the time. The association had previously been the Australian Water and Wastewater Association, which we all know now is silly. There is no such thing as waste water. All water is all useful and the same water has been around for many millions of years.

Having been out of contact with the industry since May, it was good to catch up with the gossip and the real stories that only get told over a beer. Allen was working with a group in his local Water Authority to put together a sustainability plan. Out here on the river, everything was on the table for discussion. We discussed inter-basin transfer to a town badly in need of water, the effects of that and the effects of not doing it. Passionately as I am opposed to inter-basin transfers, I had to accept Allen's observation that some environmental groups go on the attack without the relevant facts. Because it is impossible to knock their passion and commitment, it is incumbent on the water professionals to give them all of the tools they need to correctly evaluate a situation.

Allen and I paddled together on the river. It wasn't like I needed the training but it was a beautiful morning and Allen wanted to experience the water. The spare kayak was already on the grass because Jonathan had paddled six kilometres up the river to meet me on my way downstream the previous afternoon. Jonathan paddles at 8.4 kilometres an hour. Allen was much slower and was fighting the paddle. He's a fit bugger so the paddle was never going to win. It was interesting to watch what experience can add to style. Despite his lack of comfort Allen seemed to enjoy the morning.

It was a very pleasant interlude with Allen and Mary there. Perhaps Allen and I are maturing as there was not even the hint of a hangover the next day.

It seems that many people who travel the Murray like what they see in Wentworth and return to settle down. This is completely understandable. The town is quiet, the river is magnificent, and Mildura is only about half an hour away upstream on the Victorian side of the river. Most things in Wentworth seem to be an hour later than Sydney or Brisbane. This is because it is so far west in the time zone. School starts at 10.00am. We took the kayak to the Wentworth School at the end of lunch – 2.30pm. After that there was still plenty of time to do the final 12 kilometres' paddle into the Willow Bend caravan park. We called home as we watched the sun go down, a regular event for us, and Carol said it had been dark for nearly an hour.

Between Mildura and Wentworth are orchards and vineyards. The fruit growing area extends up the Darling a short way and downstream from Wentworth a short way. Not far downstream, the desert starts to hug the banks of the river. Rainfall is only 250 millimetres per year. The lush, comfortable town of Wentworth is like an oasis; down the river is harsh and dry.

A considerable amount of water was flowing over the Lock 10 weir, but this was destined for Lake Victoria in the south western corner of New South Wales. It certainly wouldn't be going down to the lower Murray in South Australia where houseboats had been stranded for months. It seemed incongruous to open and close a big lock for one small kayak so, just upstream, I pulled out of the water and walked around the lock. The lock designers don't cater for people who want to do this so at some locks it was quite a difficult exercise.

On a tree at Lock 10 is a sign that says 832, indicating that it is 832 kilometres to the mouth. Signs like this appear sporadically all the way down the Murray. The first one I saw was a 50 sign back up the Darling, meaning 50 kilometres to the junction with the Murray. At first these signs seemed absurd given the distances between everything back up the Darling. It was all part of adjusting to populated areas again. The distance to the next town, Renmark, is 251 kilometres which should have been a five day paddle. I needed an extra day though because we had promised a Koori school in Mildura that we would bring the kayak over and talk with the kids.

One challenge we didn't expect was the difficulty of getting access to the river in many places. Jonathan often encountered locked gates as he followed tracks marked on the map. After saying goodbye to Allen and Mary in Wentworth, I could only paddle 25 kilometres the next day because Jonathan couldn't get to the river after that point for many kilometres.

The wind was about 20 knots from the west, making the kayak bounce around and at times ship water into the cockpit. A skirt would have stopped the water coming in but it is restrictive. You have to unclip it to get anything out of the cockpit and that was where I kept the water bottle – between my legs.

David Bristow had sent sample bottles and eskies to us. We were to take samples, put them in an esky and courier them back to Brisbane. For any organic measurements the temperature is supposed to stay below 4 degrees and be tested within 24 hours. This sounds fine unless you know what the countryside is like. Jonathan would be more than two hours' drive on a rough dirt road from a post office much of the time. We had no ice as we used the fridge. Ice wouldn't last in an esky more than a day or so anyway, and it would need to last a minimum of three days. Hearing these protestations, David agreed not to worry about organic testing. He still got some useful results and displayed them on the Simmonds and Bristow web site. The phosphorus values were particularly interesting.

The tent was still at Wentworth and Jonathan was driving me back to it at night. We really enjoyed the place and didn't want to move until after the Koori presentation. On the second day out of Wentworth the wind was really strong. My guess was 35 knots (65 kilometres an hour) but we later found that there were gusts recorded at 100 kilometres an hour. When he picked me up Jonathan said the tent had blown flat so he had to use extra ropes and pegs and tie it to a tap next to the tent.

The wind plays a big role when paddling the Murray. It is great when it is behind you but this does not happen very often. The best to hope for is that it doesn't blow hard. When it does you try to hug the shore as much as possible. Long straights against the wind are somewhat character building but sure beat sitting in an office.

One knot, a nautical mile per hour, is about twice as fast as one kilometre per hour. The GPS on the kayak gives boat speed in kilometres per hour. Sailors quote wind speed in knots. On water with no current, white caps will start to develop at around 12 knots. At about 35 knots the tops get whipped off them. These are very rough numbers because the speed of the current makes a difference and so does the fetch, which is how far the wind has had a chance to build up waves. Travelling with the wind, the kayak made nine kilometres an hour easily. Luckily the wind comes in gusts so against it you paddle as hard as possible and hope for the best. When the gust hits, speed drops below four kilometres an hour and spray blows into the cockpit. When the gust dies down, the kayak surges to seven kilometres an hour. Lunch was in a section with the wind coming from behind, and just by holding the paddle at the right angle while eating a sandwich, the kayak would be blown along at five kilometres an hour.

On the Tuesday morning the wind had dropped. We untied our elaborate hold-down system on the tent, paid our bill at the caravan park and headed to the Koori School in Mildura. These students were well prepared. They had questions ready, having discussed the trip with their teacher and followed our progress through the web site entries. They clearly understood that we must tread lightly on the earth. Days like this lift your spirits. Knowing that you're part of a movement that can really make a difference is immensely satisfying. Speaking to the schools was undoubtedly one of the two great highlights of the trip. The other was to experience all of the tough and isolated areas with Jonathan.

The sampling eskies were supposed to be at the Wentworth post office for us to collect but after waiting for half an hour we established they wouldn't be there until the next day. Jonathan would just have to come back again.

He dropped me into the river 67 kilometres downstream of Wentworth and headed for Lock 8 where he was to set up camp for the night. This

Chapter Fifteen

sounded easy to both of us, particularly as it was only a 30 kilometres paddle. At 3.00pm I called in to see if we were within range of each other. Reception was faint and very crackly but we managed to communicate. In essence Jonathan was lost and he didn't know if he could get to Lock 8. All I had in the kayak was my jacket. Because Lock 8 would have a good road to it there was no danger that I could get stuck out on my own for the night. Or was there?

The map shows tracks coming off the Old Mail Road. Jonathan had taken Snake Lagoon Track which apparently led directly to the lock. Unfortunately, that must have been old information. There was no longer a proper track. The way in was a further eight kilometres along the Old Mail Road to what is called Lock 8 Track. This is much longer and does not look as good on the map. It may be possible to follow Snake Lagoon Track if you know the area but it seemed to Jonathan that he needed to cross Potterwallkagee Creek. He found a spot where the water was only two metres wide and very shallow but decided against crossing. If he got stuck there would be no one to help him so he opted to turn around. Dave had discovered that if you get the front and back wheels on one side of the ute stuck in the mud there is no way the engine can pull you out. Now Jonathan learned this lesson the hard way. The ground looked good. He thought it was just a flat grassy area but as he tried to turn round the left side started to sink. Efforts to accelerate through the soft mud were in vain. The wheels went down – all the way to the axle.

It took three hours to get out. Everything had to be unloaded from the trailer because one side of it was bogged too. He then manoeuvred it by hand until it was on dry ground. Then he had to dig a track for the left side wheels back to dry ground and fill it with dry dirt and sticks. When I had called him he was in the middle of this. When he had all the preparations completed he was able to drive out onto the track. The average bloke could not have done what he did. There was no chance that he could get any help. No other vehicle tracks were evident. Basically he was on his own. On the trip he had done weight training and run between 10 and 20 kilometres every day. The trailer took all his strength to move.

Having got himself unbogged and taken stock of his situation, Johnathan was back where he had been at midday, with no idea how to get to Lock 8. Another hour of driving up and down the bank did not help so he was pretty much left with no options.

The lockman at Lock 8 was on the southern bank socializing with an old guy who came there every year and set up his camp for a month. They reckoned they could tell Jonathan where to go to get to the lock. The UHF radios were out of range so I tried the satellite phone to the mobile. Luckily there was mobile coverage so I could speak to Jonathan. With the light fading, I handed the satellite phone to the lockman who told Jonathan to go back out to the Old Mail Road and drive west until he saw a landmark that he described. Jonathan did this and we called again. It took $50 in calls to talk him through the maze of tracks and out to the lock, but we did it. Half an hour after dark he arrived at the old guy's camp, covered in mud. The lockman had long gone in his tinnie as he needed to get back before it was fully dark.

We backed the trailer to the edge of the water and set up the awning. I threw a tarp over the back and prepared my bed so that my feet were less than a metre from the edge of the river. Jonathan started preparing dinner and I organised the fire. After dinner he started to recover from the tough afternoon that he had and we watched the moon rise over the river.

This was a very special night. It would never be repeated. The shadow of the earth gradually covered the moon until it was dark. Jonathan was asleep before the moon went completely dark. Eclipses happen very slowly and he is an action man. Although I had not had the stress he did during the day, the darkness put me to sleep as well. When I awoke the eclipse was over. What a day. What a night.

The wind was back the next day, about 25 knots. During the morning a houseboat set off on its journey for the day. We passed pleasantries and I noted its speed as seven kilometres an hour. Jonathan made me paddle at 8.4 kilometres an hour but that was on flat water with no wind. When we rounded the corner the wind was head on. There were three older couples on the houseboat and they seemed intent on racing me. On the straights against the wind I could hug the bank and they seemed to be buffeted to a slower speed. The only rest stops were to take a drink. Hydration is critical. It was very demanding to go almost as fast as possible but it was a lot of fun. They fell behind so that I couldn't hear their motors but every time I looked over my shoulder there they were. The pressure was relentless.

We were in an area of duckweed. This is a floating weed that chokes the river. Unlike water hyacinth you can paddle through it by keeping the speed up and just toughing it out. The houseboat would almost catch me at every section of duckweed but the sections were just short enough that I'd be out of it and away before they got to me.

There was a narrow opening in the left bank. It was not shown on the map but it looked promising. If there was a really long section of duckweed they would pass me. Taking a chance I slipped into the opening. It was much narrower than a houseboat so there was no danger they would follow. After 200 metres it opened back onto the main river. Ah ha, I had snuck a kilometre ahead of the houseboat.

Without the relentless pressure I backed off slightly. If the buffer dropped it would be game on again, but for now they were a long way behind. Ahead was the old customs house where the river boats used to pay their taxes. That seemed like a good place to stop. Looking over my shoulder there was the bloody houseboat again, as if saying 'I can wear you down.'

There was a shop at the old customs house on the South Australian border. The place had long ago changed roles, and instead of collecting taxes it now raised revenue by selling snacks and renting houseboats. I bought myself an ice cream and bottle of soft drink for lunch, unbelievable luxuries for life on the river. The manager enquired about my trip and noted the extra notch in my belt. "Been good for the waistline eh," he said. I agreed and watched my racing mates bring the houseboat in. With the 25 knot wind coming down the river they struggled to moor it but eventually were secure.

"That's great," one bloke called out. "I just made a lot of money out of you." The other guys didn't seem as enthusiastic so he had obviously taken money from all of them. I had covered that section of the river much faster than I otherwise would have, so we were both winners.

There was another bonus with the race. I had been totally focused on it so didn't have time to think about anything else. I appreciated that at the time because Ken had reappeared, in a most disturbing way. Jenny Cobbin had been sending very courteous emails to try to get the photos from Ken. We had asked her to do it because she was a lot cooler than I am and I had begun getting very annoyed. With Jenny looking after it there was no danger that I'd do anything stupid. We had taken legal advice and there had been no direct communication between Ken and me since the phone call in Bourke, and Dave (Helga) had been a witness to that. Now, as if to temper my joy at reaching the Murray, Ken sent the following email to me:

> Steve,
> Barb and I gave you the greatest gift we could have given - our time. I changed my retirement plans to help you.
> We gave up a considerable amount of money to help because we believed in the cause.
> You've used our knowledge, our teaching skills, our organisation skills, our communication skills, our people skills, our writing skills, our computer skills, our photographic and filming skills, to say nothing about our daily efforts of driving, cooking, washing, setting up and breaking camp etc. You were always talking about how much money you were going to make out of your book, the documentary, a speaking tour. None of these would be a success without our input.
> All we have asked for is fair compensation for any of our own creative work that is to be used for profit but you begrudge us even that.
> Apparently you are talking about taking us to court to take away what is rightfully our own property. I have sought legal advice and my solicitor says that the images belong to us. But life is more important to me than arguing with people such as yourself; people who have no empathy for others; people who bully, intimidate and manipulate to get their own way. I have a life to lead and I don't want to have anything more to do with you. You're not worth the grief.

K4e may have a copy of my photos. It is a cheap cost to me to be done with you.

I will still hold you to our agreement regarding any of my footage used in a documentary (a share based on the proportion of footage used) and I will expect fair payment for the use of any of my images used in the generation of any financial gain, including the use in books, articles, presentations for talks etc.

Barb thinks I am giving in to a bully but I have always believed in the message of Desiderata: "Avoid loud and aggressive persons. They are vexations to the spirit." You seem to want everything from me, well here's some advice I give you freely. Before you become a lonely old man I would suggest you reflect on your behaviour:

- Your bullying and verbal attacks on people when you don't get your own way. e.g. Your aggressive and totally unjustified attack on the young man at the camera store.
- Your attempts to make yourself look 'big' and 'tough' in front of others. e.g. Your aggressive verbal abuse of Barb and me in front of your friends at Dirranbandi.
- The motto that you are so proud of – "T.E.F.E. – Tell em! Fuck em!" It says a lot about you … and it's not good.
- The way you use people to your own ends with little regard for their rights or feelings. e.g. Your vile and pathetic attempt to bring the barman at the Hebel Hotel to tears just for the camera.
- Your hijacking of a serious environmental issue. I believe you came to see it as just an opportunity for your own glory and gain.

I think if you are honest with yourself you will not like what you see. The CD of my images will be in the mail to Jenny Cobbin as requested. Goodbye and good luck with the remainder of K4(Stev)e. (I always wondered why it was a small 'e' Earth.)

Yours sincerely,
Ken McLam

Well. Obviously the guy was pissed off. Jonathan read the email and asked if it upset me.

"Of course," I answered. "No one likes to get that sort of stuff."

There were many responses that came to mind but I refrained. Jenny Cobbin and David Bristow had both counselled me to forget it. There was also the fact that Ken's opening line was correct. They had given me the greatest gift they could. I had a slightly different idea to them on exactly what that gift was. As I saw it, Jonathan wouldn't have been on the trip but for them. The people who had been bypassed for Ken and Barb would have been there. Jonathan came to help his Dad out in a tight spot and it turned into an adventure we will both remember as long as we live. However indirectly, I have them to thank for that rare father and son opportunity.

Still, I don't ever willingly back down from a fight. It was difficult to do what my friends advised and ignore the email.

That afternoon Jonathan met me at Lock 6 and we picked a camp spot just back down the road and around the bend out of the wind. It seemed that the wind was picking up. There was no grass on the ground so we had a small fire to cook dinner and then made sure it was out. Jonathan had erected his three man tent for us. It has lots of vents with mesh and an overall fly that covers the tent. Because it was really warm this was a clever idea. We just lay on our sleeping bags rather than climb into them and swelter.

The wind had been strong earlier. Now it started blowing fiercely. Some of the gusts must have been about 50 knots. The trees would roar as a gust came through. When the gust hit the tent we would be sandblasted by dirt blasting in through the vents. The vents do not have covers so the best we could do to avoid the blast of stinging particles was position our sleeping bags so that our heads were out of the direct line of fire.

One big gust hit about midnight. The whole tent went flat with one of the curved fibreglass support sticks hitting Jonathan on the head. The gust had woken me so I watched the tent collapse onto him. It woke him up and he pushed the stick away which made the tent spring back into shape. I started laughing. Despite being so uncomfortable it was really funny to think how much more uncomfortable Jonathan was.

Chapter Fifteen

The last time I had enjoyed that feeling was on a bike trip with a mate in 1969. It was in the days before full face helmets. My jacket was what was called a blanket coat as we couldn't afford anything decent. It was pissing down rain and we were soaked and freezing. The rain was stinging my face so I had a handkerchief over it to try to stop the stinging. It was soaked of course. Breathing sucked the wet hanky into my mouth and I almost choked. I laughed because I thought about what a shit of a time my mate would be having. Maybe I need a therapist.

We awoke to cold and breezy conditions. The front had passed and it would clear up and be a reasonable day. It was also an exciting day because we were to meet another sponsor in Renmark that night. This was Dave Alexander, Australasian Manager for the pump company KSB, his son Kurt, and Mottie, the KSB South Australian manager.

My log for the day says:

> **Steve:** *Another stroke of luck! Yesterday I was heading north west and the wind was from the north west. Today I am heading south east and the wind is from the south, south east. Ah well, I was told this would happen.*

Well it doesn't say what kind of luck had struck. Luck does not always have to be good.

I was now well into South Australia and heading for the town of Renmark. At Renmark the river takes a big turn back east before heading south and west again. Standing on the southern bank of the river in Renmark it is disconcerting because you know the river is heading west but here it is flowing directly east through the town. The concept had always been difficult for me when driving through but coming down the river it all made perfect sense.

Chapter Sixteen

Renmark – Waikerie

It was Saturday and we were having a rest day. My hands were sore from the really hard paddling against the wind. Dave, Kurt and Mottie arrived from Adelaide about lunch time. They brought a letter of encouragement from the Young Water Professionals in Sydney which lifted my spirits enormously. We had an obligatory beer and discussed accommodation options. There were no cabins available at the caravan park because of a major sporting carnival on in town. It was either a top class motel for them or a weekend roughing it in the tents with us. Kurt, Dave's 10 year old son, opted for the tent and his elders agreed.

We had Jonathan's three man tent, a spare sleeping bag, a self inflating mattress, a cover and a pillow. Dave went shopping for enough extra gear to make them comfortable. "We need some extra fart sacks," he explained. Dave's family are campers so the extra sleeping bags would get used. With the shopping expedition successfully wrapped up we prepared the kayaks for a paddle through Lock 5. Dave and Kurt were in the double one, and Mottie was in the spare single kayak. Jonathan was cameraman for the day.

Jonathan and I helped the visitors down the bank and into the river without tipping them over. They were clearly excited to be on the water. Kurt had a huge grin on his face the whole time, Dave was having a ball, and Mottie was concentrating intensely on staying upright. Even pottering along slowly we covered a fair bit of ground and it wasn't long until Lock 5 was visible.

Mottie struggled with his balance until he got in front of Dave and Kurt, then his intensity increased remarkably He was determined to make it to the lock ahead of them no matter what. That's probably the sort of spirit that makes him a state manager. It was amusing to watch, and he didn't relax until he won his self-styled race. Jonathan arranged for the lock to be opened and we paddled in. This was the first lock that I had paddled through, although I had walked around five. The big gates closed behind us and the water started dropping. We were all excited by the drama of the occasion as the water dropped right down. The tiny kayaks were dwarfed by the massive steel and concrete structure. From being near the top we dropped way down. Our perspective was still from the water level so it felt like the sides rose up around us. The lock is probably three metres high. The gates opened at the other end, and we paddled out onto a different landscape with more sand and higher banks.

We talked with Jonathan on the UHF radio and arranged a pick up about four kilometres downstream. For blokes who looked uncomfortable at the start, the guys were paddling along really well and were quite confident. Kurt in particular seemed to be loving it. I considered the value of organising weekends with parents and their kids paddling kayaks and camping together. There couldn't be many better things to do in life than that.

Only two kayaks will fit on the roof of the ute and there is just a squeezy little place on the back seat for a third person because of all the gear packed in there. Jonathan drove Mottie and Kurt back with their kayaks and Dave and I waited. We've known each other through AWA for many years and have shared many similar ideas. Dave is one of the nicest blokes you will ever meet but don't cross him. He's not a 'turn the other cheek' man, either.

Before we left the caravan park I had showed Dave the email from Ken. "No," he said. "That's not you. I know you and that's not you."

We talked about the effect of the email. He understood perfectly my desire to respond, my feeling that I wanted to fight. But like the rest of the team, he counselled me to do nothing. That was Dave, Jenny and David, all people whom I respected, telling me not to respond. So that was that. There would be no response. If the photos were still not returned then I'd go on the attack after the trip finished, but if they arrived that would be the end of it.

My talk with Dave was the turning point. Since arriving at the conclusion that Ken and Barb had stolen the images the whole sequence of events had been eating me up. Everyone told me to forget it and I knew they were right. When you feel wronged though, rationality does not always win out. Gradually, over the next couple of weeks, I thought about the issue less frequently and finally forgot it altogether.

Jonathan arrived back at the river so we hoisted the kayak onto the roof and drove back to camp in Renmark where we set our chairs up to watch the sun go down. You can't really say it is romantic sitting there with a mate, glass of red wine in hand, watching the sun set over the water but there was a certain contentment, an appreciation of the beauty, of friends, and experiences shared.

Dave was paying for dinner at the club. The walk was about two kilometres but no one thought that we would drive there. Consensus ruled on the meals too. There were no thoughts that the smaller steaks should be considered either. It was big ones all round.

Some unkind family members have said I snore. In fact the family member sharing the tent with me thought my snoring was a great joke. However, my skills in this regard were proven to be pathetic. The noise I make is nothing compared with what came out of the other tent. Kurt was tired and went straight to sleep so he missed the racket. Jonathan and I were five metres away, and we reckoned it made our tent vibrate. The noise went on all night, sometimes dropping to a deep rumble, sometimes reaching an ear splitting crescendo. It was all part of the fun.

It was Father's Day and Kurt had bought his Dad a mug. We had four mugs and there were five of us so it couldn't have been a more appropriate present. Breakfast complete, our three visitors set off back to Adelaide. Jonathan and I went for a drive to see the Southern Ocean. He wouldn't be with me when I eventually reached the sea, but I wanted him to see it so we could discuss strategy. Because he is a keen surfer his considered opinion would be valuable. We had seen each other beaten up by the Pacific Ocean at times and we both knew the difference between discomfort and danger.

The call of the Southern Ocean was back again, stronger than ever. From one great ocean to another via a great inland river system – we were coming to the climax. I was excited but apprehensive at the same time. The sea is different in different places. Knowing the winds, the tides, the moods of the sea at home were not going to be a lot of use 3000 kilometres away on a coastline facing Antarctica. Local knowledge can often mean the difference between life and death. it was the drama of this challenge that I think was drawing me, and the excitement of paddling on an ocean so vastly different from the one I knew.

The round trip was 800 kilometres and it took all day. We didn't get to the Murray mouth but we looked at the beach at Goolwa and decided if that had to be the entry point, the tactics to be used when I got there would depend on the conditions of the day. The beach gradient was flatter than we were used to on the east coast. This meant that there were twice as many waves to get through between the sand and the calm water. Picking a break would be twice as hard under normal conditions, and with a decent swell it could be impossible.

West of the Murray mouth, the coast from Victor Harbour to Cape Jervis is quite isolated with many cliffs and steep beaches. The distance through the surf zone is much shorter there but the break is strong. The size of the waves makes it practically impossible to get out to sea from any of the beaches. They are big solid buggers with a lot of water in them. If something went wrong and I had to make a landing, there would be no easy way to come into the beach through that surf. I'd have to paddle fast behind one wave, staying in front of the next one as long as possible. When it got to me it would tip me over and I'd swim in with the waves washing the kayak to the beach. I wasn't too worried about doing all that, but getting back out would be impossible.

My plan to paddle all the way round to Cape Jervis meant accepting the high probability that if I had to come in to shore for some reason it would be messy and there would be no chance of getting back out to sea again. Mobile phone coverage was very patchy along there as well, so I'd be pretty much on my own.

We left the area understanding more about the coastline and sharing the same thoughts. The practicality of paddling this leg would depend on the weather and there was a distinct possibility that the Southern Ocean would tip me over and spit me out.

It was a great way to spend Father's Day.

On Monday morning we had an interview with ABC Riverland. It was in the studio and Jonathan came with me. Being his first ever radio interview he was uncomfortable with what he had said and kept thinking of things he could have said better. There was no reason for him to worry. The ABC later mailed us a copy and everyone who listened or heard it reckoned he had spoken very well and conveyed his thoughts articulately. It was another special occasion to cherish.

We made two interesting contacts from the interview. A bloke with decades of rowing experience called us to talk about changes in the river and his marathon rowing events for charity. These days he doesn't do much more than 25 kilometres at a time but at 80 years old that seems pretty fair. Liz Frankel, an artist from down the river near Waikerie, called in with an offer of accommodation at her place. Being more than 100 river kilometres downstream from Renmark, it would take us a few days to get there but the invitation was to prove very convenient.

We also covered two school presentations and made some panel repairs on the ute, so didn't get to the river until 1.00pm. Because we had just been to the Southern Ocean, I talked to the kids about how to keep water out of the kayak. You wear a spray deck which is tight around your waist and clips over the edge of the cockpit. It is also called a skirt. There were lots of cheers to encourage me to show them what a kayaker wearing a skirt looks like. The kids might have been taking the piss out of us but it was all good fun.

The sun was disappearing over the horizon again by the time I met Jonathan at Berri. The next morning we had a memorable presentation with the Riverland Special School. One of the students had Tourette's Syndrome which caused her to shout out. Watching how the others handled this and also watching Jonathan's reaction was quite humbling. You could feel the support that everyone was silently giving this kid. It also had a funny side. The wombat says "…the animals, the water, the sky, the ground, the bugs, the fish, the tacos, the people: they're all connected. Everything is connected. They all depend on one another and if you ignore that you're doomed. Repeat, doomed. So listen up. It's one world, not two worlds, not three worlds…"

When he got to the word 'tacos' there were howls of laughter. We usually show the film twice as most kids like to see him again. We thought that second time round they might hear what he said after 'tacos', but no. When the wombat said it, peals of laughter rang out just as before. Jonathan and I looked at each other and couldn't help laughing along with them. We were both very glad we had made the effort to stop there.

Arriving at the river at 10.30am meant a quick sprint to Lock 4, about nine kilometres away. We had arranged to see a local kayaker, also called Steve, who would paddle from there to Loxton with us. He and Jonathan were waiting when I arrived. There was a boat ramp upstream and a track past the lock down to a sandy beach. This was an easy portage and with only three more locks to go the end of the river was in sight.

The other Steve set a brisk pace but the sea kayak was faster so if I waited back for a photo I could catch up again after a few minutes of working flat out. Steve talked about his favourite camping spots along the river and they were pretty impressive. He doesn't go to any of them at Easter though as there are far too many people. It was about 25 kilometres to our camp site and as we paddled Steve talked about salt and salt interception.

Salt is a huge problem throughout the entire river basin and I was beginning to understand what it is all about. Too much water applied on the surface carries the salt from the soil as it drains down to the river. In many places the height of the land above the river is 100 metres so the salty water easily drains down such a steep gradient. One method used to prevent this salty water polluting the river is to dig cut-off trenches at the bottom of the escarpment to capture the salty water, so that it can be pumped away to evaporate.

Loxton is the southernmost town in the loop of the river between Renmark and Morgan, the heart of South Australia's Riverland. This area has been heavily irrigated to grow citrus, grapes and stone fruit. Huge tanks of wine abound. The story of the degradation of the land opposite Loxton is sad. It is now a useless salt pan. It is heartbreaking to see such a wonderful resource as the Murray exploited through ignorance or greed and being poisoned for generations to come.

Here is some of what my fellow kayaker had to say:

> *There are two sections along that stretch of the river where there has been significant environmental degradation due to salinity caused by poor irrigation practices in the past. Both of these sections are claypans that sit between the river and the cliffs. Run-off and seepage from over-irrigation have caused a rise in the level of groundwater in these areas which has brought saline groundwater into the root zones of the redgum and blackbox trees.*
>
> *There's a misconception that the river is sick because of a lack of water. The long term degradation of the river and the floodplains isn't due to a lack of water as such but rather an over regulation of the flow which has changed the natural wetting and drying patterns that are necessary for the survival of many of the local species. Maintaining an artificially high 'pool level' between the weirs on the river has drowned many formerly healthy floodplains.*

Chapter Sixteen

This is most evident in the (former) evaporation basin at the southern end of Katarapko Island, directly opposite the Loxton township. Stormwater and wastewater from the local winery and fruit juice factories were too saline to dispose of directly into the river so a series of levees was constructed around Yabbie Creek and the Horseshoe lagoons on Katarapko Island to create an evaporation basin where this 'waste' water was stored until the next high river 'flushed' it out.

Needless to say, highly saline water destroyed most of the redgums that were unfortunate enough to be inside the basin. I had written a letter to the 'appropriate' government minister in the early eighties about this issue and was told that the evaporation basin was a good thing because it created a habitat for water birds. Yeah, I shook my head in disbelief too! Thankfully we've progressed slightly since then.

The evaporation basin levee banks were dug out in a few strategic spots a few years back and the water allowed to drain out of the creek and horseshoe system. I go for a walk a couple of times a year along Yabbie Creek (where I used to go for a paddle), it's good to see things slowly restoring themselves. Despite our past attempts to well and truly fuck things up Mother Nature has a way of restoring the balance.

Let's hope it's not too late when it comes to climate change.

I should also add that there's been a lot of good work done in recent times to correct past mistakes; I think we're slowly learning. Salt interception schemes are 'intercepting' highly saline groundwater before it reaches the river and channelling it to already saline areas such as the Noora Basin to the east of Loxton.

There's also a been a lot of research into creating an artificial wetting and drying cycle in some of the world heritage wetland areas at Chowilla, between Renmark and the borders of New South Wales and Victoria. These trials have had great results in the small areas they've been tested in. But what's the point of having a few hectares of 'restored' habitat surrounded by a wasteland?

We arrived at Loxton caravan park at 4.00pm and pulled the kayaks up to the tent, only to have the local newspaper reporter ask us to put them back in the water for a photo shoot. Steve's mate also arrived to take him and his kayak home. He's not a keen kayaker like Steve and I can understand why. He only has one hand. That doesn't stop him holding a drink in his good one and telling a decent yarn though.

At the Loxton School the next morning we enthused some of the kids sufficiently to send us regular emails as we travelled along for the next 100 kilometres or so. Two memorable guys call themselves the 'Loxton Legends' and thought everything was 'awesome'. They were full of life and reminded me of Jonathan not that many years ago. One of the guys said that he was not going to wash his hand after he shook mine, so I put on the web site that he had shaken the one that I hadn't washed since Brisbane. Maybe that changed his mind.

We were camped next to the water and because the caravan park is so huge we had that area to ourselves. It was a powered site so the computers were plugged in. There was also the added bonus that it was free. The woman who owned the park had heard us on the radio and wanted to support the trip so she wouldn't accept any money. Bless her, and the ABC, which was responsible for spreading our message.

The plan was to stay two days. Jonathan would come and pick me up for the second night and then after that it was off to Liz Frankel's place where we would leave the trailer. It was almost time for Jonathan to return to Brisbane, so a safe haven for the trailer was much appreciated. As we headed into the more populated regions closer to Adelaide the road and the landscape became very busy. The highway was beside the river at Kingston-on-Murray and the roar of cars and trucks seemed out of place with the river. Plenty of pumps hummed away and some of the pump stations were huge. Many people were out and about in tinnies on the river trying to catch fish, despite the fact that the cod season had just finished. No one seemed to have caught anything but it seems that's not why people fish.

Climate change issues were being explored more fully in discussions, as people were not so defensive down here. People generally accepted it as a fact, not a theory. There is no doubt, anyone dry land farming was having a lot of trouble and even those with irrigation had seen their allocations slashed, so most were doing it tough and waiting for things to improve.

The situation was worse than anyone could remember. With the frontal systems that bring the rain now more inclined to pass through below Tasmania, people understood that this drought may have something to do with climate change. Most people like to relate what they are being told and more rain on the Southern Ocean was not much use to them.

Jonathan picked me up at 49.1 kilometres for the day which left just 37 kilometres and Lock 3 to negotiate the next day. Then we would store everything with Liz before we headed off for home. We left the camp in the morning before the sun came up. By 7.30am the sample bottles for Simmonds and Bristow had been filled and I was away from Kingston with Liz's house in my sights. Liz's instructions were easy to follow. The pace was fast and furious that day as I knew there would be no more paddling for a few days.

Although we hadn't met Liz, we had chatted on the phone. She and her husband Clint run a studio making glass art. Liz writes as well and Clint teaches to supplement their income. Their house is high up on the banks of the Murray River and she had given Jonathan instructions on how to get there to meet up with me. Built with recycled windows, running on tank water and generally conforming to as many sustainability practices as possible, the house had me intrigued. As did Liz, who is pretty much in despair about her beloved Murray River and what we have done to it. Greed is what she puts it down to.

Originally the plan was for Jonathan to drive back to Brisbane and swap with John Crocker who was to finish the trip off with me. That would have allowed me to continue on my way until John turned up. Because it is a very long drive for both of them to do on their own, it made more sense for me to make the trip and share the driving in both directions.

When he arrived at Liz's, Jonathan had a problem. The kayaks on the trailer had hit the rear window on the canopy and broken it. He had tried to find a replacement with no success. "Not to worry," I said. "Let's just make sure everything in the back is secure and we will get it fixed in Brisbane." We confirmed with the canopy supplier back home, Opposite Lock, that they had one available so it was probably convenient that we were going to Brisbane anyway. Liz gave us a copy of her book for school children and we were gone.

Now that the time had come, both of us were anxious to get home. It was about twenty hours counting stops and we shared the driving all the way. We have one golden rule that must not be broken. If the driver gets tired he stops. This can be at any time. Once, the rule was applied after ten minutes' driving, and after half an hour a few times. You just never know when it will hit you so the trick is to know the signs and stop, no matter what.

Back in Brisbane Opposite Lock fitted a new glass and wouldn't accept payment. The support that we were receiving from all quarters was quite humbling.

For Jonathan and me it was the end of our time together. We posted the following on the web site but also sent each other more personal and private messages.

> *Steve:* Now, before I let Jonathan say anything I want to reflect on the time we had together. For me it has been great. He has been rock solid in some pretty difficult circumstances and I really can't express just how proud I am of him. I can't believe how much we ate some times and the food was always great. We also covered a huge distance from near Bourke to South Australia and saw some fantastic country together. Good luck on your round the world trip and don't forget to send photos especially of the surfing spots in Mexico. (I don't envy you surfing in Canada, you can have that to yourself.) Finally, to have had you helping out for the past seven weeks is something I will certainly treasure for the rest of my life.
>
> *Jonathan:* After an interesting drive back with Steve to complete the change over of support crew, I arrived back in Karalee on Friday at

Chapter Sixteen

midday. Looking back on the trip, Steve and I shared some quality time and I got to experience some beautiful scenery in Australia's amazing outback. Along with getting the chance to support Steve and witness him successfully carrying out his journey I also learnt about two of the most famous rivers in Australia and the dreadful state they are currently in. Furthermore, it was enjoyable to see Steve convey his message of sustainability to not only the farmers and everyone he met along the way, but also the children at the schools he spoke at.

Being Steve's son and knowing his strong and determined character, I always knew he would complete this journey successfully once he set off. He is now at the tail end of the trip and the finish is in sight. If you have been monitoring Steve's progress on the website, be sure to keep a close eye on things from here as the finish will come around surprisingly quickly.

Best of luck to John and Kareen Crocker with their support crew stint. To Dad, "KEEP ON TRUCKIN". You have come a long way and the finish is now in sight. It was a pleasure to be involved in the trip and I have no doubt you will succeed to the finish.

CHAPTER SEVENTEEN

The Last of the Murray

THE BREAK IN BRISBANE HAD BEEN VERY WELCOME. There was repair work to be done and the support team to see. We had arranged a team lunch on Sunday and it was important to let them know that provided Ken sent the photos as he had agreed, the problem was now behind us. I had shared his email with them and made some comments on it so that was off my chest. Jonathan understood just how fast things might unfold with the last legs of the trip but we still had many schools to see, and depending on how many Jenny arranged, the end date was still in doubt. There were the vagaries of the Southern Ocean to consider as well, so it was impossible for anyone to book flights to be in Adelaide at the end of the trip.

Saturday and Sunday flew past. Carol and I said goodbye for the last time during the trip. Jonathan and I said goodbye for many months and maybe years. On Monday morning I picked up John Crocker early and we drove all day. John was listed on the web site as the manufacturer of the kayak wheel system. Geoff had just updated the site to include him as a support crew member as well.

John is in his 60s but is an excellent boilermaker and quite fit from the work that he does. We met at Aquatec in 1985 but between 1994 and 2003 when he came to Watergates, had gone our separate ways. At Watergates his experience was invaluable and we made a great team. I'd do the engineering designs but he would tell me what could be made

Chapter Seventeen

and what the potential problems were. The design and manufacture of the wheel system for the kayak were a part of this process. It is probably the mutual respect we have for each other that would forge the bond for a successful trip relationship.

We stopped at Hay for the night and were in time for lunch at Liz's on Tuesday. Driving through Mildura reminded me of our almond lots. When we sold Watergates I continued to work and built up some savings. Our accountant said that he had just the place for the money and it was very tax effective, meaning the government reduced tax on your earnings if they went into these almond trees. "It is somewhere in Tasmania," advised the accountant. But it wasn't. The trees were near Mildura using Murray River water. The government had effectively paid me to invest in something that was sucking the life out of the river. Although it was a shock to learn this, there was not much that could be done. We were committed. At least the trees are drip irrigated. Some say they require much less water than citrus trees but it seems to me that both require the equivalent of about one metre of rainfall per year which is four times what the natural rainfall is. What we have are large areas of citrus trees being bulldozed while new plantations of other trees and grape vines are springing up further and further away from the river. Is this not madness?

Arriving at Liz's just after lunch gave us more time this visit to see some of her work. Her passion for the environment and understanding of the plight of her river are conveyed in children's books and a moving slide show of birds and animals with the background music 'What About Me'. It certainly was fortuitous that she had rung the ABC after she heard Jonathan and me on the radio. She reckoned that we had sounded 'fair dinkum and intelligent.' Phew, at least we impressed someone.

When we pulled the kayak down to the river we noticed that there was a flat tyre and a broken weld but as John is a qualified boilermaker that didn't create any problems. We simply changed the wheel and John looked after it. He kept saying how much things had changed since he was last in the area. I was interested in what he had noticed but no wonder there were changes – he hadn't been back since 1972!

Tuesday was a cold day. Liz Frankel's place was only 20 kilometres upstream from Waikerie but the wind was head-on all the way there. Luckily the journey was only that long. Until now the water had always felt cold but that day it felt warm. As it is unlikely the water temperature changed, the wind must have been pretty cold. Liz had told us to look for salt running out of the cliffs. It was not hard to find. Little stalactites, a few centimetres long, were all over the cliff face and drops of salty water plopped under the overhangs. Looking at a cliff from the water you can see the difference between where water continues to drip and where it has stopped. The wet areas are brown and the older, dry areas are closer to white. Manoeuvring the kayak under the overhang to take a picture was like moving into a rain shower. No wonder the Murray is salty.

At the Waikerie caravan park John was his usual sociable self and talked to all and sundry. One of the grey nomad women recognised him from the web site, his first taste of fame puffing out his chest. He was also starting to get the trailer and ute sorted out to his liking. Admittedly he did grumble about our general cleanliness when he found some knives in the trailer covered in peanut butter and dust. John and Jonathan had worked together at Watergates and continually ribbed each other so John was enjoying the opportunity to revive old rivalries. They had each made parts of the stainless steel frame over the trailer. One of John's sections had broken near Wilcannia so they made comments at the other's expense and relayed these back to the Watergates workshop. Over time though, there was no doubt that the over-sixty bloke got the better of the twenty-one year old. You just can't beat old age and cunning.

From Waikerie to Morgan was a long but easy day. Morgan is where the river turns southwards for its final journey to Lake Alexandrina and the Southern Ocean. It is here that the benchmark EC or salinity measure is taken. Fruit trees and vineyards can tolerate up to 800EC. The level at Morgan was 800EC. With zero room to move, the anxiety of growers must be extreme.

There was a surprise at Morgan. A huge pump station is constructed on the northern bank. It pumps water from here to Whyalla on the opposite side of Spencer Gulf, a distance of over 300 kilometres. The river Murray supplies 90 percent of the water for South Australians, most of

Chapter Seventeen

whom do not live in the Murray system. The inter-basin transfer is huge. It is staggering to think that water that once would find its way down the Snowy River to Bass Strait can now find itself halfway across Australia in the opposite direction.

It seems that we are now heavily into water re-use. Mostly that means we take water out of the river for our towns and cities, treat it to three star or maybe four star standard and then use it to irrigate trees. Sounds good in principle unless you are a river relying on that water for your life. We have the technology, so why not treat it to five or six star standard and put it back in the river, *upstream* of where it was taken to make sure local water authorities follow the rules. I have no doubt that such ideas will become common place eventually, but when? How much time do we have? All we seem to do is take. The water in the Murray is important for its health. It shouldn't be taken 300 kilometres away and it shouldn't be extracted for urban use and not put back.

We took more samples for Simmonds and Bristow to evaluate but because of the difficulties of keeping them cool over long distances, they again didn't include biological analysis. Inorganic testing is still important, and it includes phosphorus, probably the main precipitator of algal blooms. Tests for phosphorus showed that it was above recommended limits in specific locations. To speculate on why this occurs is dangerous, but it is puzzling that in a very severe drought it would happen. Often a major manmade source can be runoff from adjacent properties but in a drought it would be unlikely that this would occur from rainfall and equally unlikely that anyone would over irrigate.

The tent was still set up at Waikerie caravan park where John had met his first fan. We had a meeting organised at Liz's place nearby for Friday afternoon with Karlene Maywald, South Australia's Minister for the Murray River, so had decided to keep our base at Waikerie. Because the river headed north west from there and then turned south it wasn't too much of a problem to come back to the camp site from all the way down river to Swan Reach.

Lock 1 is at Blanchetown. This was a milestone because it meant we were now following the natural level of the river. All ten locks had now been negotiated so there would be no more man-made steps in the height of the river. Going through the lock just on closing time I felt the Southern Ocean getting close. Between this lock and the ocean would be Lake Alexandrina. Contemplating that was exciting.

The river south of Morgan is characterized by lots of cliffs. These are often seen in magazines and on postcards. They are spectacular but one gets used to them on the river. Coming round one bend is very similar to previous bends. South of Blanchetown the view became very special. The wind had been howling overhead and I hugged the western bank. To the east were high cliffs. Storm clouds were above and to the south and east but the sky was clearing to the west. As the sun dipped low in the western sky, it shone on the cliffs turning them bright gold. With the dark clouds as a backdrop the view was stunning. The roar of the wind in the trees added to the drama. No one else was around to see it. Despite the cold, the potential for a drenching and the discomfort of the wind, to use one of Jonathan's words, it was 'awesome'.

On Friday morning as I paddled towards Swan Reach there was a racket on the road way up above the river. It sounded like a lot of heavy vehicles but the noise approached very slowly. When it got close enough I could see that it was a group of tractors. They stopped and waved and I could hear them yelling so I paddled across the river to them. Unfortunately we were still separated by high cliffs and couldn't hear each other. I could see figures but couldn't make out any details. Eventually we had a final wave as we departed in our separate directions. 'What a strange group,' I thought.

John picked me up early as we had Karlene and the Waikerie School to visit. "You will never guess who I saw this morning," he said. "Did you see any tractors on the road?"

"Yes, but I couldn't see much. They stopped and yelled something but it was all a bit strange."

"They met you in Dirranbandi on their way to a tractor show in Queensland."

Chapter Seventeen

I was flabbergasted. It was Marj and Sam. John said they were on their way to the tractor show near Liz's place. We drove to Waikerie school, did our usual presentation which was in the hall with the whole school and had an hour to spare before our meeting with Karlene. We had time to go to the tractor show. Driving around there were hundreds of antique tractors. Eventually I said, "I know that one."

Marj was there with a big hug and it was handshakes all round from Sam and their friends. The Biloela trip had been a great experience for them. They had kept tabs on my trip via the local media but had expected to be away at the very time I'd be in Mannum where they lived. They were excited about seeing me on the river, meeting John at Swan Reach and now actually catching up for real. I still can't work out why Marj likes bouncing around the country with Sam at 30 kilometres an hour in a contraption that is so noisy you cannot hear each other even if you shout. Then again, she probably wonders about a bloke dragging a wheeled kayak along for thousands of kilometres.

It was time to get to Liz's to see the minister. We needn't have worried about timing as she was late. She also had a family commitment later so we crammed the meeting into a lot less than half an hour.

Karlene Maywald is a very interesting case. The South Australian Government is Labor. Karlene is a minister in that government – the Minister for the Murray River. She's also a National Party member. She's a bundle of energy with a deep understanding of the dire plight of the river. She went through graphs of just how bad things were but noted that finally people were starting to understand.

She talked about a new desalination plant to get water for Adelaide.

"That is a good move," I said, "provided you use renewable energy to power it and you're careful with the design of your diffusion mechanism for the concentrate."

She seemed to agree but detailed discussion revealed that the renewable energy wouldn't be additional energy. It would be from a wind or other renewable energy project that has already been committed. This is where we diverged. There is no point claiming that you're using renewable energy if you add an energy consuming device to the grid and do not add a renewable energy source to provide that power. This is the crux of the concept of 'additionality'.

Concerned about the cost of this, Karlene asked, "What about the pensioners?"

With no time to debate the issue, the meeting was over. I understood that she meant low income people have a right to clean water but I believe there are plenty of ways of providing that without wrecking the environment.

Another way to look at the price of urban water is to think of its value and how it fits with other goods and services. One of the blokes at an AWA committee meeting said that in his family their daily expenses were: $1.50 for water, $3.50 for electricity and $15 for information technology (being phones, internet, television and the like). Without water there is no life. Why should a whole tonne of it only cost you about a dollar?

In any case, South Australia takes a dividend from SA Water of 20 percent. When you think about this, it is a staggering conflict of interests. Our most precious resource is a cash cow for governments right around the country. Many of them survive on the revenue raised from the sale of water. On one hand governments are trying to teach people to use less water, on the other their revenue is tied to the amount that they use. How's that for smart?

We left feeling that the seriousness of the crisis had been understood by the minister, that she was gradually convincing others of this, but that very few people were actually connecting the dots on sustainability. This was further demonstrated by what we saw down the river, where despite the dire consequences staring people in the face, sprinklers were watering grass in caravan parks.

Climatologists talk of tipping points. From what I had seen we've been exploiting the land and the rivers since around 1860. The crisis on the Murray is severe. It may never recover. We are faced with the dreadful choice of deciding what to save and what to let go. Have we crossed a tipping point? Did we do that a hundred years ago when the swampy land with plentiful tree cover was changed to dry grasslands? I suspect that we have, and the only answer is to try to get the land as close to natural as possible and then learn to work with nature, not against it. We need rivers to work as rivers. The natural systems must be restored and we must work with nature to do this. Only then will we have a productive and sustainable environment to provide our food for future generations

South Australia is highly dependent on the Murray River. With much of the population living outside the catchment of the river this involves a huge inter-basin transfer. With modern technologies it is also unnecessary. It is possible to take water from the Southern Ocean, desalinate it and pump it up to Adelaide using only renewable energy. At Adelaide it is easy to recycle the water. There is a storage area right in the city called the Torrens River which is currently just a drain for much of its length.

Karlene Maywald may understand the seriousness of the crisis but understanding on its own is useless. Adelaide should prepare itself to take zero water from the Murray. More of the same thinking will get more of the same results – more crises. A paradigm shift is required. Is there anyone in power prepared for this? I doubt it.

Brisbane people have responded extremely well to the call to reduce water consumption. It is half that of most other Australian cities. How can Brisbane do this when the situation in South Australia is probably dire compared with that of Brisbane? It is called public education. People were able to change their habits and when they did, they realised that they didn't really need all that town water anyway.

It took a while to pack up the tent and drive to our start point the next day so it was not until 10.00am that the kayak touched the water. Despite the late start I clocked up 55 kilometres to get to Bowhill. There was a houseboat going the same way at about eight kilometres per hour so that kept me amused for three hours. He didn't win the race but did toot his horn a lot. In the end he stopped for the night and let me continue. At Bowhill there was a houseboat full of retired Kiwis so there was plenty of rugby talk and banter.

The day hadn't been unusual. Sure I had a couple of beers talking to the Kiwis that night but that was also fairly normal. At 10.00pm I woke feeling really sick and went outside to vomit. This is a very rare thing for me. It continued during the night. In the morning I went to the toilet at the usual time but the action was practically a non-event. Initially, the only effect of the illness was that I decided that paddling was out of the question. By 9.00am though, John took me to Mannum hospital where things deteriorated.

Sitting on a surgery bed spewing for the umpteenth time was not a lot of fun. The nurse said, "That's brown, it must have faecal matter in it." 'Bloody great,' I thought. 'Now I'm spewing up shit.' My real worry was that I might have a bowel obstruction. This had occurred before, about 15 years ago, and it is very unpleasant. The signs were similar. They talked of an ambulance trip to Adelaide. It would just be laparoscopic surgery but it might be a week before I'd be paddling again.

Eventually I was checked by a doctor. He prescribed exactly what was needed: Maxalon to stop the vomiting, morphine to reduce the pain, and a saline drip to combat the dehydration. He didn't think it was a bowel obstruction. He thought it had been caused by dehydration. The fact that I had covered about 3000 kilometres successfully without dehydrating didn't sway him. For the whole trip I had been very careful with fluid intake and watched my urine colour all the time so I was absolutely sure that was not the cause of the problem. It was obvious that I was dehydrated from everything going out and nothing going in. That was all we really needed to know.

By late afternoon I could hold down a few mouthfuls of water and by that night I knew that everything would be fine. A teenager was placed in the bed next to me with what were similar symptoms to mine. Now the diagnosis seemed clear to me. He and I had the same virus. This may have been over simplistic, but it was logical to me. The other explanation was a temporary bowel blockage that fixed itself after relief from the drugs.

The next morning I asked to be discharged. This caused some consternation with the medicos but eventually a senior doctor agreed and away I went. Walking back into town I did feel a bit weak but by the time I got to the caravan park my spirits were high. Mannum Hospital had been fantastic. They had fixed the problem and they had been wonderful to deal with. When I had the bowel obstruction in the early 90s, Cairns Base Hospital had operated quickly and fixed the problem. When I almost karked it on the Plenty Highway the air ambulance and Alice Springs Hospital did a great job. Overall, it seems that the hospital system around the country does an excellent job on acute care.

While I was in hospital John had his own problems to deal with. The tent at Bowhill had blown flat. Winds of over 100 kilometres an hour had caused some damage to the rods and he had struggled to get the lot down and packed away. After this effort he somehow managed to erect it again in Mannum and go to Adelaide to pick up Kareen, his daughter. Kareen would be with us until the end of the trip.

The Monday was fine and calm. Kareen was at the tent with John when I arrived from the hospital and we had a relaxing day stocking up from the supermarket, a pub lunch and a snooze in the afternoon. The sprinklers were on at the caravan park for a few hours. The locals considered this normal. The river level was low. It looked like the tide was out except that it is not tidal. The sign at the amenities block said that they have a licence to use the water and that none of it comes from the town water supply. Funny thing though, the town water supply comes from the river, the irrigation water comes from the river, and the river is running out of water. What is it that these people do not understand?

A year later and they seem to have 'got it'. The river level is lower, one ferry cannot operate and there is no water at all between the caravan park and the island. The sprinkling has stopped though, and the toilet block has waterless urinals. This is just one reminder that I received on the trip that people have to hurt before they comprehend.

Tuesday morning Kareen came with us back to Bowhill to resume the paddle at 7.30am. We arranged for a radio contact at 1.00pm. The morning went well and at 12.15pm while John and Kareen sat in the sun at the tent I walked in dragging the kayak. John was surprised to see me and had something to say about our agreement that I should take things a bit easy as it was the first full day out of hospital. I was feeling fine and said so. We would continue at least for another couple of hours. Unfortunately, the next logical place to finish for the day did not turn out to be that logical. Ultimately I met John and the local press after dark at Murray Bridge. It was a 73 kilometres day.

Is this madness? Was it perhaps just a silly thing to do? Well, it had no ill effect on me. It was great for motivation though. A mere tummy bug was not going to interrupt this trip. Over the past four days we had covered 128 kilometres which was an average of 32 kilometres per day and during this time I had spent a bit over 24 hours in hospital. To me, that was pretty satisfying and a good sign that I could surmount any difficulties that might lie ahead on Lake Alexandrina and the Southern Ocean.

Wednesday started fine but quickly deteriorated. The wind came up and the waves were big enough to make paddling quite wet. Weather was starting to be an issue. In the past we had watched the weather reports but apart from rain forcing us out of an area impassable in the wet, we had just coped with whatever the weather dished up. With a lake crossing and an ocean looming, it became a whole lot more important.

John and Kareen drove into Adelaide while I made my way to Wellington, the last town on the Murray before the lake. They collected Nell, who had seen us on the 7.30 Report and wanted to come along as support. She would only be with us for a week before the trip finished but like the rest of us, Nell had a great time and experienced much. She had her own tent, I slept in the back of the ute, and John and Kareen slept in the big tent.

From Wellington we drove back to Tailem Bend for a school presentation. That was when the rain squalls started. We hoped they would be over by the time we went outside but that was wishful thinking. It was still squalling at the end of the day when I pulled in at the last farm before the lake.

The forecast for Thursday was for moderate westerly winds. Typically the route that day was due west. Even so, it could have been worse. It could have been strong westerly winds or gale force westerly winds. For Friday, light northerly winds were expected. Saturday could be anything. We had come to learn that the weather coming off the Great Australian Bight changes quickly and so predicting more than two days out could be risky.

The Southern Ocean was my big worry and that would be the governing factor in what we did. Bashing against the wind and the waves on Lake Alexandrina would be tough but hey, this was supposed to be an adventure wasn't it? There had been a couple of emails from experienced kayakers who had said the lake could be dangerous. One recommended that I terminate at Wellington. One recommended that I travel around the shore. Another had cautioned me about the Southern Ocean and recommended that I terminate at the Murray mouth. These were all things to think about but I made my decision. The spray deck would be fitted, I would wear the life jacket with the Emergency Position Indicating Radio Beacon (EPIRB) in the chest pocket, tie the paddle to the front of the kayak and keep the stern rope on so that I could hang onto the kayak if I ended up in the water.

My plan was to hug the lee shore of the finger of land where the river meets the lake, and then go straight across the lake to Point Sturt. The crew seemed apprehensive when I set out but I was looking forward to the challenge. The 73 kilometre day had lifted my confidence and I was feeling strong and fit.

For the first hour it was just like paddling on a very wide river. I was pleased to have the shelter of the finger delta as the wind was quite strong, gusting to about 25 knots but mostly blowing around 15-18 knots. I had sandwiches and a big drink in the lee of the tip of the finger delta and then poked my nose around the corner and set off. The GPS had Goolwa as an optional way point which I had selected because it was in line with Point Sturt. This gave me a bearing to the landfall that I wanted to make. These waters were not exactly familiar territory and Point Sturt was below the horizon so this gave me some comfort.

The wind dropped slightly so that it averaged about 15 knots. The gusts dropped the boat speed to as low as three kilometres per hour but it got as high as seven kilometres per hour when the gust passed and I was still paddling at full strength. After an hour a smudge appeared on the horizon. 'Maybe that is where I'm going,' I thought. Another two hours and it really did look like land in the position that it should be. By then the wind had dropped to about 10 knots so I had more to eat and continued in high spirits. For fluids I kept a bottle outside the spray deck where I could grab a drink at any time.

Eventually we made radio contact. John was very relieved. As I approached land, the ferocity of the waves decreased. I noticed a channel marker and near that was a kayak. Nell had taken the spare single and paddled out about a kilometre to meet me. There was not a spare spray deck so she was getting well and truly wet. To us Queenslanders the water was very cold but Nell is from Tasmania and she seemed to be enjoying herself.

You can see the southern shore all the way across. Without a GPS or compass you could use that as a reference point, assuming that conditions allow you to see that far. I doubt that anyone should contemplate a trip across the middle without navigational aids. There was not a soul around on the lake proper and it was wet and uncomfortable against the wind. At no stage did I feel in danger but the crossing was certainly an adventure and gave me a great sense of accomplishment.

As predicted, the wind did swing overnight. In the morning it was about 12 knots from the north which was gentler than the previous day and was from behind. With such a long fetch the waves are a lot faster. In the rivers I paddle faster than the waves but today they rolled underneath me. My top speed was 10.4 kilometres per hour and I was mostly achieving around 8.7, so the kayak was flying. Actually it was more like flying by the seat of the pants because I was only guessing where to go. Despite that I ended up in exactly the right position. This was con-

Chapter Seventeen

firmed by stopping at a farm house not far from the barrage at the end of the lakes, and asking where I was.

At the barrage I made radio contact with John and we figured out where to meet. It was to be on the Coorong opposite where it joined the sea. There were speed boats zooming about the Coorong and there was a village on the island. John held the UHF to the phone in the car and I had a patchy conversation with Amanda. This was a bit emotional. The great river trip was done, from Brisbane to the Southern Ocean. What had seemed surreal just a few months ago was now a done deal. More importantly it was only a little over a year since I had been in a great deal of trouble, lapsing in and out of consciousness on the Plenty Highway.

I could not believe it. The Murray River does not even reach the sea, it stops at a man made wall, below sea level. I wrote in the first chapter about a barrage across the Brisbane River to collect fresh water, mainly as a way to get people thinking. Here was one across the end of the Murray River that failed to keep the water fresh.

Think about a river system the size of the Murray Darling. Its rivers run for thousands of kilometres. The water from the mountains of the Great Divide on the east coast has travelled down those rivers, through a lake, then into the Coorong and out to sea for millions of years. But that was before we got clever. The Murray River now flows into Lake Alexandrina which ends in a dam wall that we call a barrage, separating the lakes from the Coorong. Statistics are difficult and open to interpretation but since construction, chances of a modest flow to the ocean of, say, 150,000 megalitres each month have fallen by 60 percent. Modest flows now occur less than half the time they used to. We have kept a lot of water. Some we've diverted hundreds of kilometres away, some we've pumped onto the land, washing salt from the earth to poison the river. This is the greatest river system in Australia and we have stuffed it up.

As I mentioned above, the water upstream of the barrage between Lake Alexandrina and the Coorog looks entirely different from that downstream. Downstream it is like it is at the beach. Upstream is a dirty colour with a telltale brown that indicates that the water in the lakes is salty. The farmers there had suffered as their fresh water turned brackish and increasingly salty. The barrages were built around 1940, along with the thirteen locks up the river. Recently the water level had been pumped so low that it was below the sea level and salty water is seeping back into the lake. More troubling is presence in this area of what is known as acid sulphate soils.

These are all over the place on the coastal rivers of New South Wales and Queensland. When Watergates supplied a gate to a council in New South Wales we found that the creek that they were trying to fix had a pH of 2, the same as battery acid. This sort of thing could happen to vast areas of of the lower Murray River, Lake Albert and Lake Alexandrina. Such acidification simply cannot be allowed to occur.

On the other side of the barrage is the Coorong. The Coorong is a very long, thin section of water parallel with, and very close to, the coast. It used to be linked to the lakes and had unique salt and freshwater tolerant species adapted to the major changes in salinity. With the construction of the barrages, the lakes were to stay fresh and the Coorong would be salty except for water released from gates in the barrages. That was the plan anyway. Once a wondrous place, it is dying and it is people who are killing it. Dredging is being carried out to try to save the Coorong. It is at what they call the Murray mouth. You cannot drive to this area. There is not a road. Perhaps we want to keep the Murray mouth a secret because it is not the mouth of a river. It is an opening in the land where the Coorong is attached to the sea. Our iconic river finishes ignominiously at a man made wall.

When I set out from Brisbane it was with an open mind. My major goal was to learn about this great river system. At the end of it I was humbled, saddened and ashamed. It was people like me, civil engineers, who had wrought this destruction. We had been so proud of our achievements, of conquering nature and sculpting the landscape to our desires.

The parts of Australia where I travelled were once wet. They were cool swamps with rivers forming just part of the water landscape. There

Chapter Seventeen

were springs where water flowing through the land sometimes seeped to the surface. The springs have gone. The swamps have gone. They have been replaced by giant evaporation ponds connected by channels that dwarf the rivers. Though I have paddled the length of many rivers, describing what a river is still eludes me. The complexity is similar to that of veins and arteries with their electrolytes, red blood cells, white blood cells and a myriad of other components. A river's function is no less important and its influence extends well into the oceans. Our very existence depends on the health of our rivers but as we degrade them and the landscape we gradually condemn ourselves to an uncertain future.

Very few of our dam builders, hydrologists, engineers, irrigators, or even whole water authorities, understand what a river is. They talk about sustainable yields but they do their sums by looking at what happens when rain falls, how long it takes to flow somewhere, and how much you can keep. In other words they make drainage calculations. How sad is it when people at Surat would prefer to have no water at all from the Darling Downs than the toxic slop that they had been getting?

John and the crew were waiting. We agreed that they would drive back through Goolwa and come down the beach to the mouth while I paddled the few kilometres around the Coorong to meet them.

The sea had been calling but I was becoming worried. For the last few kilometres the call had become a pronounced roar. It was now continuous. There were no discrete waves. From this, and the level of the noise, it was easy to deduce that the waves would certainly not be small.

I had completed the journey down the river, but now had to confront the ocean.

CHAPTER EIGHTEEN

The Southern Ocean – Adelaide

THE CREW TOOK A LONG TIME to get to the meeting point on the beach. This allowed me time to study the bar. It seemed that the way out to sea was down the western bank of the river mouth, paddle out behind the waves breaking on the beach, wait in a deep section for a break and then head out to sea at an angle to the east. Waiting and watching is essential to getting through surf. After twenty minutes watching I figured it would all be fine. The lull between the waves was enough to get out. From the beach to safety was about 700 metres and it would need patience and care. Once outside though, it would be an easy paddle of less than 30 kilometres to Victor Harbor.

When the crew arrived I was keen to get underway but I wanted it all filmed and photographed so I took the time to show Kareen where I'd be going and where she needed to be with the camera.

This was almost familiar territory. The sea is something I understand. It is to be respected but not feared. It is unpredictable but if you watch for long enough you can get a fair idea of what it might do. Unfortunately the wind had just swung from the north to the north east, so it was blowing along the shore. This was not a big deal but the waves did not stand up as long as they had been, so were breaking further out.

Chapter Eighteen

'Patience,' I thought as I waited in the gutter for the right moment for the assault. 'Now! Let's go.' I moved at full pace to spend the shortest time possible in the danger zone. Two hundred metres later and I thought I might have nailed it. 'All the waves are behind me, just another hundred metres and I'll be past any evidence of broken waves.' Always optimistic, I powered on as fast as I could go.

'Oh no!' A hundred metres in front of me a huge wave appeared and broke into a wall of white water 1.5m high. 'Go for it. Crash through,' I thought. But a sea kayak with wheels is not streamlined to crash through waves. The wall of white water pushed me backwards two hundred metres before I extricated myself. The kayak had taken a lot of water through the spray deck. It didn't steer properly and it was low in the water. I was also over to the east in the wave zone. 'Never give up. Never fucking give up. Paddle like you have never paddled before,' I said to myself. And so I did. Another big one came through, sent me backwards, tipped the kayak over and the paddle rope broke.

The kayak was washed away about a hundred metres. I went to it as fast as I could. I got there, just before another wave came. I tried to tip it back over but the sea wasn't having any of that. It stripped the kayak away and left me with just the paddle again.

The beach was a bloody long way off. 'Shit!'

Plan B: swim into the beach, walk along it to find the kayak, swim out and retrieve it, get it back to the beach, drag it back to the mouth, cross the other side, then go and have a beer and think about another strategy.

And so a long swim began. I was disappointed. It had been so close, so very close. But that is the sea. It can be cruel at times. It can also be cold and I'd much rather have been doing this in Queensland. 'Bloody southerners, how do they cope with water like this? Oh, that's right. Most of them don't go in the water. That is where great white sharks live.' These were pretty stupid thoughts and were discarded immediately.

After a few hundred metres I was elated. There was my hat! And there was the kayak! Unbelievably they were only about 30 metres to the west. I could see my water bottle there too. It was all in a gutter about 200 metres off the eastern beach. Swimming over to it, I grabbed my hat 'Yee ha! This is good stuff.' I was cold but not that cold.

I tried to turn the kayak back over. It wouldn't go. I tried with my feet on the wheel, again with no luck. It took a long time before it righted. Then I had to sort out the various bits of rope all over the place. The string on the emergency beacon was wrapped around my neck, the life jacket and the paddle. It was a struggle to make sense of all this and get it sorted out. My brain seemed a bit fuzzy and none of my limbs responded as they were instructed. Eventually the mess was cleaned up enough to slide into the kayak. There was no hope of baling it out but it did paddle even full of water. I retreated, chastened by nature for my boldness.

Unfortunately the crew didn't have experience with surfing like Jonathan and me. They thought that I had been safely out behind the waves and so headed off in the ute. On the way they stopped at a sandhill to watch but couldn't see me. A couple of rangers told them what had happened and so they quickly drove back along the beach.

John and Nell grabbed the kayak and Kareen took my shirt off while she rubbed me with a towel as I started shivering violently. We emptied the kayak, put it on the roof and headed back to Goolwa. The kayak could be fixed but neither the UHF nor the VHF radio would work. To top it off the camera I had worn around my neck since Dalby was broken.

There were no photos and no footage of my troubles but I was past caring. Given the troubles with Ken and his insistence on wanting his share of profits the whole matter of images seemed far too much trouble anyway. Goolwa Masts and Marine welded the rudder back together. John and I threaded control cabling through from the pedals and we discussed the events of the day.

Chapter Eighteen

With the kayak now repaired we really needed local knowledge so we went to the information centre in Victor Harbor. They were very helpful and gave us the sea rescue number which turned out to be the local police. The local police explained that the system was they would ring the sea rescue guys who would contact me. Ultimately a Senior Constable at the Water Police in Adelaide called. He was very helpful, gave us information about the forecast winds and we had a good general discussion about the waters in the area.

Despite his helpful advice, I missed having volunteers on the ground. At home, I'd just wander down to the local sea rescue station, get the low down and make a well informed decision with the volunteers watching the same weather as me.

Without a radio it would be irresponsible to set out around the peninsula. On top of all that, weather conditions were too unfavourable to make the attempt anyway. It would be crazy to set out on a nice calm morning and be bashed by a 30 knot westerly in the afternoon. The other factor was the tide. To get to Cape Jervis and have to paddle against a five knot tide wouldn't work when my sustainable top speed is just over four knots.

Reluctantly, I decided to give up my cherished hopes of paddling the Southern Ocean. The only course of action was to walk over the Fleurieu Peninsula to Gulf St Vincent. The team back in Brisbane was of the opinion that it wouldn't lessen the journey and the only problem was in my head. After half an hour I just accepted the situation and cheered up enough to sleep like a log.

Maybe it was because I was near the end. Maybe it was the pleasant day. For the first time my thoughts drifted to what the trip had been about. It really had been a great adventure. Unfortunately it had been seriously marred by Ken and Barb and I wondered how I'd have handled things under different circumstances. When you're at your physical limit, issues like this become more difficult. I hadn't realised that their agenda was not simply to support me to ensure that I made it. It had taken months for me to see this. The signs were all there. In the end when Ken wrote such a nasty email I understood that there were people around whose personal agenda mattered more to them than the big picture. This was new to me. Never had I seen such a display of vehement immaturity, but there it was.

Perhaps there was another way to look at it, though. Rod had suggested that to complete the trip I just needed to take one step at a time, and that is how we are to beat global warming – one step at a time. This was a good analogy except that the government seemed to be going in the wrong direction, which would be like me walking to Rockhampton before turning around and heading for Adelaide. Perhaps the Ken and Barb issue is one way to understand the challenge with living sustainably. No matter how great the challenge, some people will put their own motives ahead of every thing else. Perhaps the scale of what is happening is just too awful to accept. But accept it we will. Eventually, when it is in our face we will just have to accept the reality that the world will change in some very big ways.

The wind on Saturday was 35 knots from the north and the swell had grown. Putting the kayak back in the water at the Murray mouth I was relieved that I didn't need to have another crack at the bar and headed west along the Coorong towards Goolwa. The crew were at the barrage about an hour later and we took the last of our water samples for Simmonds and Bristow. There was a beach on both sides and a bitumen road so walking around the barrage was easy. A short but cold and wet paddle later and we pulled the kayak out at Goolwa. The high winds had some sail boarders out and they clattered their way past me as the boards whisked along the tops of the waves.

Although it was cold on the water, the 20 kilometres walk to Victor Harbor was hot. John bought a water bottle to replace the one that I had lost at sea but at Port Elliot I stopped to indulge myself with an ice cream. I was enjoying walking in civilization. I could just stop at a shop and fill my tummy with whatever I liked.

Chapter Eighteen

The hills across the middle of the peninsula looked daunting. It was 37 kilometres over land to Normanville. The weather report was for lighter northerly winds all day Sunday, switching to southerly on Monday. This would be perfect. It would blow me up the Gulf towards Adelaide.

As the sun came up on Sunday I headed off towards the hills because I reckoned it would take all day to get over them. The weather was perfect, warm but not too hot. Down in the valley there was no wind. The grass was green, a huge change from what I had seen for the last four months. Even the hills didn't bother me.

The road ahead was downhill for nearly a kilometre. I wondered whether it would be possible to ride the kayak down it. The paddle was in the cockpit so I tied it to the harness frame, sat on the bow and gave it a try. With my feet as brakes I could control the speed. With the paddle I could more or less control direction. There were no cars around so away I went. If someone had seen what I was doing they probably would have had me locked up but it was great fun. The hill ended and as the centre of the peninsula was still ahead the road started climbing again.

The next hill was the big one. After that it would be all downhill. At the top the wind was blowing from the south. The weather system was nearly a day early. What did that mean for tomorrow? The only certain thing was that it would be uncertain so there was just one option. When I reached the coast I'd need to start paddling again.

Nell was at the top of the hill with the video camera. "I have a trick," I said. "Hop in." She looked at me like I was a crazed lunatic. "It is easy," I said. "Watch this." After tying the paddle to the harness frame it seemed like the arrangement was a little better than before. Nell had moved a few hundred metres down the hill so I rolled down to her to allow her get some footage. The trick was to keep my feet on the ground to slow the forward motion. To lift them would mean certain disaster.

Nell will no doubt remember the trip down the hill for the rest of her life. Hopefully she will recall it as exuberant fun, but maybe not. It was all going perfectly when the road widened for 100 metres where it was possible to overtake. The steering was far from perfect and there was a considerable delay in turning. We drifted to the middle of the road and a silly woman drove up behind us. As we were bringing things under control and moving to the left, she was trying to overtake on the left, despite the fact that there had only ever been half a lane there. She slammed on her brakes, tooted her horn, and then went around to the right as she should have done in the first place. As I said: silly woman.

It was roughly 2.00pm when I reached the beach at Normanville. The wind was about 12-15 knots from the south, making the sea a bit lumpy and creating enough shore break to make the start a wet one. Because the wheels are just behind the cockpit, their supports catch waves and splash them down my back. After a bit of a dousing it was off northwards to Sellicks Beach some 20 kilometres away.

Initially there was a beach to my right but it stopped abruptly at large cliffs. There was enough swell to make the waves disconcerting. Riding with the incoming swell was fine. When it hit the cliff and bounced back it wasn't too bad. It was when the two sets of waves collided that I felt vulnerable. There were no boats and no people around. There were no beaches to get into shore. If the worst happened I could always swim to a rock, plan an exit onto the cliff base, climb the cliff and walk for help. This was not very appealing so I paddled out another 500 metres and the going was much better.

Pulling into Sellicks Beach as the sun went down I was tired. The day had involved 37 kilometres of hilly country dragging the kayak and then 20 kilometres of paddling in the Gulf. The paddling had been a bit stressful as the darkness closed in towards the end, and I was not sure where I had to aim for. Glenelg Beach at Adelaide was now less than a day away. After a pub dinner back at Normanville sleep came quickly.

We planned to stay at a caravan park ten minutes' drive north of Glenelg. John was not familiar with Adelaide and he had no street

CHAPTER EIGHTEEN

directory but he managed to get there. The agreement was to meet at Glenelg at about 2.00pm.

The trip up the Gulf was stunning. There was no wind at all for most of the trip. It is always difficult to make out low landforms from so low in the water but my instincts had started to work and I managed to stay in a straight line and not get caught behind sand bars and rocks. There were a few board riders out but given the small swell they earned full marks for optimism. It was disappointing that the camera had broken as there were many photo opportunities.

A few hundred metres ahead there was a fin cruising in the water. My course would miss it by about 50 metres so I just watched it slowly moving through the water. 'Big fin,' I thought. Getting closer there was another fin near it but a lot smaller. What to do? To go closer to the shore was to go into an area where a wave could tip me over. In the end curiosity got the better of me. After all I had six metres of bright yellow plastic to sit in. I paddled towards the fin. It saw me, turned over and disappeared. It was a seal. Maybe it had just been lolling around in the water with its flipper in the air waiting to scare the living daylights out of me. If so the plan had worked.

Glenelg Beach appeared and I paddled the kayak onto the sand. John had been delayed setting up the tent and Kareen and Nell had decided to do some domestic duties anyway. I dragged the kayak up the beach and a group of young Japanese tourists helped me lift it up the steps onto the pavement. Although soaking wet I ventured into the pub and bought a beer and a glass of champagne. When John arrived we had another drink to celebrate the occasion. The journey was all over bar the shouting. Technically this was part of Adelaide. It had been a long trip.

The next morning we recreated the beach arrival for the television, met the South Australian president of the Australian Water Association and headed into Adelaide with the morning traffic. Amanda arrived from Brisbane with five kilometres to go and we walked together. She was the only person from Brisbane. Carol was away in Italy with friends on a trip that had been planned two years ago. She had worried that she might be away when the trip finished but I persuaded her that I couldn't finish before the end of October. Now, I'd be back in Brisbane in time to pick her up from the airport. Rod and Shirley would be here for the weekend but that would be too late. We would be back home by then.

Adelaide was buzzing. One of their teams was in the Australian Football League grand final on the weekend. Buzzing is probably not strong enough. It was only Tuesday and they were going nuts. This would continue all week until the game, and if they were to win the real party would start.

Amidst this excitement Amanda and I arrived at City Hall. The mayor greeted us with a gift of a book on Adelaide. I presented him with the letter I had carried from the Brisbane Lord Mayor and John cracked open a bottle of champagne. There were lots of interviews which are now all a blur. We loaded the kayak onto the ute and that was that.

Cameramen had asked us to show jubilation, to jump for joy. That was not how I felt. It had been a long trip with many milestones on the way. Reaching the Murray with Jonathan was a very big milestone. Entering the Southern Ocean had been momentous. Here I was, over 2,000 kilometres by road from home and home was where I wanted to be. We had covered a total of 3,250 kilometres. Paddling accounted for 2,170 kilometres and walking 1,080 kilometres.

Back in Brisbane there were some media interviews and stories and then there were articles to write for technical magazines. Sure, I had proved to myself that I was tough. I felt immense pride in completing such a journey when things looked so bad after the motor bike accident. But it is what I learned about what makes a river system that changed me. It had been such a long way, so many puzzles to try to comprehend. But I was finally starting to makes sense of how water works in the whole landscape. Even if only for that, the trip was worthwhile.

Epilogue

RETURNING TO SOCIETY after such an extraordinary experience standing back and looking at it objectively is very difficult. The madness continues. We eat up finite resources and change the planet in ways that threaten our very survival. Most people simply do not connect the dots.

The worst thing we can do to our atmosphere is to extract and burn fossil fuels, and yet we are increasing the number of coal mines as fast as we can. We are also madly building the infrastructure required to export it to other countries so they can burn it. To do this, we pay people a premium. Every person who makes money from coal, or who lives off money from any part of the coal industry, is stealing from future generations. But they do not see it.

The river system I saw was definitely in crisis. Reducing extractions to re-establish an environmental flow will not be enough. Almost all of the natural springs are gone, wetlands have all but disappeared, there are vast tracts where nothing grows because water tables have dropped. These wetlands were the living tissue that nurtured the landscape, a sponge that absorbed water and released it during dry times. We have ripped the sponge asunder and exposed its heart to the relentless sun.

Many people were in denial and just waiting for the cycle to change. But it is far more serious than this. The damaged landscape is reflected in social failure and economic failure. Irrigation is supposed to bring wealth to an area. If this is the case, why do Bourke and Brewarrina look like towns under siege? Why have their once vibrant shopping centres declined to almost nothing? If irrigation is so good, why does irrigation cause such divisions in communities? Why are the non-irrigators happier than the irrigators?

I'm ashamed to be part of a society that has wreaked such devastation. That Young Farmer of the Year 2005, Graham Finlayson, has the right vision for the future. With his organic farming methods he's making his land more productive. If others could do this, imagine the communities that could develop, the work that could be provided to supply the infrastructure for all the successful small farmers.

We need to follow his lead. It's time to say that we've got it wrong. It's time to try again, in a more natural, sustainable way that rewards building up the land resource for the future, not grabbing what we can and hoping for the best. It's time to recognise that the long term management of the natural systems is a necessary condition of our survival.

Many people call me an environmentalist, or a conservationist, but I do not see myself as either. I am just a civil engineer who happens to be able to read the writing on the wall. I experienced the antagonism of some rural people toward perceived 'greenies' but most of us are on the same side. We all want sustainable, thriving, bush communities. We all want to build a way of life that benefits our children and grandchildren.

If what we build is not sustainable, we have robbed them of their inheritance. From my observations, that is exactly what we have done.

Our river systems are precious. If they die, we die. And they *are* dying.

Index

4WD..............123, 154, 170, 181, 197
Achilles tendon 47, 74, 90, 99, 110
acid sulphate......................*See* acidity
acidity...............................26, 33, 255
Adelaide...248
Al Gore 38, 163
Alexander, Dave 17, 228
algal bloom........................66, 85, 194
almond trees.................................242
Andrews, Peter 77
Aquatec....................................80, 241
aquifer 50, 73
Artesian Basin...............................154
Australian Water Association
 (AWA)...7, 16, 18, 30, 55, 130, 218, 230
Back from the Brink.........................77
Balonne Minor.......................119, 134
Balonne River 101, 106, 119
barbed wire.........................68, 98, 160
barrage.........................9, 254, 261
Barwon River........................119, 169
Beardmore Dam....................106, 119
bifurcation 126
biofuel...204
Birrie River 119, 134, 156
Black Betty.......................................14
Blanchetown 245
blister..150
boat speed......................................221
Boating Camping Fishing (BCF)...18, 22, 44, 163
bogged..................21, 154, 197, 222
Bokhara Hutz.................................155

Bokhara River...............119, 150, 153
Bourke 174
Bowhill...249
breakdowns 114, 123, 150, 205
Bremer River...................................35
Brewarrina 159
Brisbane...19
Brisbane River 5, 9
Bristow, David 18, 147, 189, 220
Broken Hill.....................................199
Brown, Peter........................10, 18, 93
Buchan, Robert.....111, 115, 120, 171
bull shark ....*See* sharks
bulldust..204
Bundamba STP 36
callistemons....................................74
carbon dioxide..........................6, 47
 sequestration 47
 vehicle emissions 15
Carter, Bob....................................164
Catchment Management
 Authority..................................202
cattle...............92, 101, 103, 135, 153
cattle grid ...12, 95, 130, 139, 193, 198
characters
 Allan Murdock 174
 Allen...218
 Amanda...........*See* Posselt:Amanda
 Bil Stallman 93, 97
 Bill (with bridge).......................63
 Bill Gorman............................120
 Bryce.....................*See* Jones, Bryce
 Carol...................*See* Posselt:Carol
 Cathy Finlayson 155

INDEX

Chris (Hebel) 139
Darrin (physio) 14
Dave Alexander.......*See* Alexander, Dave
David (surgeon) 13
David Bristow....*See* Bristow, David
Davo 12
Dermot 187
Di *See* Thorley, Di
Dick at Dalby 77,79
Don Alcock 114
Elizabeth.....................................95
Farnie.....18, 37, 59, 104, 111, 167
Gail McBride 16
Geoff...........................10, 16, 113
Graham Finlayson....................155
Greg (Chinchilla) 80
Greg Hoadley 101
Heather.....................................159
Heidi 11
Hugo 188
Jenifer ....*See* Simpson, Jenifer
Jenny ....*See* Cobbin, Jenny
Jenny McDonnell 25, 27, 50
John (Stanthorpe) 136
John Crocker .....*See* Crocker, John
Jonathan ....*See* Posselt: Jonathan
Karen 136
Kev Flanagan..............................55
Kurt 228
Liz ....*See* Frankel, Liz
Malcolm (River Lady) 209
Marj and Sam McClelland ..125, 246
Marty (farmer) 125, 135
Mary 218
Monty....................*See* Ryan, Dave
Moses 161
Mottie 228
Murray......................................194
Nell...251
Nina (Distance Teacher) 178

Ollie..188
Pat (River Lady) 209
Pat Stallman 93
Peter ....*See* Brown, Peter
Robert Buchan...*See* Buchan, Robert
Rod 204
Rory and James 112, 114
Shirley 130
Steve (drover)...........................103
Steve (kayaker) 234
Tammy.......................................80
Terry Loos ....*See* Loos, Terry
Tom 127
Tony (Kogan pub) 85
Uncle Bill...................................71
Vikki ....*See* Uhlmann, Vikki
Wally Mitchell.........................177
Whale, Sperm............................32
Charlie's Creek...........................81
chemical pollution 85, 99, 103
chlorine......................................58
Clarence River.................26, 136, 177
claypans...............153, 158, 186, 195
climate change
   contrarian...............................164
   denial.....................................177
   models.....................................56
   observations 77, 186, 248
   presentation 19, 20, 38, 48, 176
   skeptic.....................77, 172, 164
Climate Swindle.........................163
coal 6, 46, 78, 86
Cobbin, Jenny 17, 46, 76, 89, 109
Condamine River 78, 84
conductivity...............................33
contour mounds.........................63
coolibah tree.....................153, 195
copyright 113
cotton 97, 120, 132, 142, 178
Courier Mail...............................80
Crocker, John 15, 142, 241

269

Index

Crocker, Kareen 250
Cubbie Station 110, 132
Culgoa River 119, 126, 133, 169
Cullen, Professor Peter 106
Daily Examiner 26
Dalby 75
dams ... 8, 26, 208, *See also* ring tanks
Darling Downs 65, 89
Darling River 169, 171, 214
Davis, Chris (CEO AWA) 30
Department of Natural Resources
 and Water 118
Department of Primary Industries .. 110
desalination plant 246
Dirranbandi 121, 125
dirt roads 69, 181, 196
Distance Education Centre 178
dogs 75, 139
Doyle, John 115
drainage .... *See* water management
drought 38, 42, 56, 60, 102, 133,
 153, 171, 188, 238
ducks 5, 208
dummy spit 25, 53, 94, 138
Dying Darling 55
Econnect 16
El Niño 56
emus 13
Epco 18
EPIRB 252
evaporation 171, 236
extraction ... 76, 81, 94, 119, 133, 243
farm costs 84, 152, 157
feet problems 53, 78, 109, 150, 201
Finlayson, Graham 155
fish 91, 94, 99, 176, 191
fish ladder 161, 191
fish traps 160, 190
Fisheries Department 176
Flannery, Tim 6, 82, 115, 177
flat tyre .... *See* breakdowns

Fleurieu Peninsula 260
floodplain .... *See* overland flow
Frankel, Liz 233, 238, 242
frost 77, 97
fuel consumption 15, 140
Gatton 45
Glenelg Beach 263
Glenmorgan 96
Global Positioning System .... *See* GPS
goanna 205
golden oldies 92, 116
Goodooga 149
Goolwa 232
Gorman, Bill (MDBA) 120
Gowrie Creek 66
GPS 23, 32, 88, 92
Grafton 26, 177
Great Dividing Range 31, 39, 41,
 83, 136
greenhouse gas ..... *See* carbon dioxide
Greenhouse Mafia 164
greenie 84, 95, 115, 178
grey nomads 77, 160, 202, 243
Griffith University 31
Gulf St Vincent 260
Hamilton, Clive 164
Harbison, Michael 265
Hash House Harriers 31, 145, 151,
 159
Hay 242
Healthy Waterways 36
Hebel 139
helicopter 90
historical account 60, 63, 91, 94,
 153, 158, 235
Hoadley, Greg 101
holistic farming 77, 101, 110, 157,
 195
Hood, David 39, 78
horses 47, 103
horticulture 43, 60, 115, 136

INDEX

Hospital Creek 159
houseboats 209
   racing 224
hydrogen 140
Idalia 187
Inconvenient Truth, An 163
Ipswich 19, 39
Jandra 188
Jeanes, Susan 39
Jondaryan 70
Jones, Bryce 22, 27, 30, 43, 50, 111
Junction, The 35
K1 racing kayak 72
Kapunda Park 117
Kayak4earth 16, 189
Keytext 111, 114
Kogan 84
KSB 17, 228
La Niña 56, 71, 118
Lake Wetherill 206
land management .... *See* holistic farming
Laws, John 164
lignum 153
lock 217
   Lock 1 245
   Lock 4 234
   Lock 8 222
   Lock 10 219
Lockyer Valley 43
Loos, Terry 62, 93, 118
Louth 71, 187
Loxton 236
magic mile 66
maps 22, 92, 170
massacre 160
Mayors
   Adelaide ...... *See* Harbison, Michael
   Bourke .... *See* Mitchell, Wally,
      See O'Malley, Wayne
   Brisbane .... *See* Newman, Campbell

Ipswich .... *See* Pisasale, Paul
St George .... *See* Buchan, Robert
Surat .... *See* Stewart, Donna
Toowoomba .... *See* Thorley, Di
Maywald, Karlene (SA MP) .. 246, 248
MC^2 16
media
   balance 164
   behaviour 80, 87
Menindee Lakes 206
Minister for the Murray *See* Maywald, Karlene
mitchell grass 154
Mitchell, Wally 177, 181
Moles, Sarah 55
Morgan 243
Mt Crosby 8
Murray Bridge 251
Murray Cod 191
Murray Darling Basin Commission 120
Murray mouth 232, 255
Murray River 111, 136, 208, 217, 248, 254
Narran Lakes 118
Narran River 119
Natural Capitalism 156
Newman, Campbell 29
NSW Institute of Technology 26
O'Malley, Wayne 169
Oakey 69
oats 97
one armed paddler 233
Opposite Lock 239
orchards 60, 136, 178, 219, 242
organic farming ... *See* holistic farming
overland flow ... 94, 102, 132, 133, 158
Ozgreen 17
ozone 58
Ozwater 5
pain 88, 193, 249

Index

pasture....77, 101, 152, 154, 157, 204
pelicans..5, 208
pH..*See* acidity
phosphorus.....................66, 220, 244
pigs...129
Pisasale, Paul 39
Point Sturt 252
politicians...........42, 48, 82, 177, 246
Pooncarie......................................211
port (wine) 117, 189
Posselt
   Amanda......11, 114, 194, 254, 264
   Carol 10, 78, 141, 241
   family..10
   Heidi...11
   Jonathan........11, 15, 178, 233, 239
probe...31, 36
pump........72, 76, 135, 169, 237, 243
Queensland Natural Resources and
   Water..135
Queensland University...................45
rain.....67, 76, 84, 121, 123, 126, 196
rainfall records 63, 83, 153, 158
Rangeville School 62
recycled water.......*See* water, recycled
renewable energy...............6, 204, 247
Renmark 229
rice..178
ring tanks.....72, 73, 75, 94, 121, 149
Riverland Special School 234
Rose Isle..176
Rosenhek, Ruth 17, 106
Royal Flying Doctor Service...........11
rugby 116, 249
rugby league 141
Ryan, Dave (Monty) 13, 32, 145,
   149, 179
salt..........33, 171, 182, 195, 235, 243
Schluter, John 90
Scorcher..164
Sellicks Beach 263

sequestration 6
sewage treatment........8, 36, 66, 136
shags..208
sharks....................................5, 36, 258
sheep...188
shoulder problems............13, 68, 193
Simmonds and Bristow ..18, 220, 244
Simpson, Jenifer 8, 55, 62
snake......................................67, 205
solar panels...............................29, 44
sorghum.................................84, 152
spray irrigator 76
spring....................*See* water sources
St George 111, 153
St Mary's School............................62
Stallman ......*See* characters, Pat & Bil
star rating ....*See* water, rating
Stern, Sir Nicholas..........................48
Stewart, Donna 106
Stone, Professor Roger 56, 61, 71
stormwater harvesting.....................9
Stumer, Lloyd 38
Sunmap..22
Sunwater.......................................118
Surat 105
swamps ....*See* water sources
Swan Reach...................................245
Tailem Bend..................................251
TEFE 48, 105, 226
telephone..............23, 104, 111, 167
   bills......................................65, 247
   coverage 34, 141, 183, 223, 233
Telstra................24, 87, 167, 183, 199
temperature.....*See* water temperature
testing......................*See* water rating
Thorley, Di.......................58, 60, 136
Tintinalogy 205
Toowoomba 19, 41
Total Dissolved Oxygen..................33
Total Dissolved Solids....................33
towns..............................86, 175, 199

trailer design 20
Traveston Dam 9
trees 61, 100, 105, 158
Trevallyn 194, 201
Turnbull, Malcolm 58
Two Men in a Tinnie 82, 115
UHF 23, 59, 104
Uhlmann, Vikki 19, 38, 107, 169
ultraviolet light 58
United Utilities 18, 22, 27
University of Southern Queensland 56
Van Wisse, Tammy 214
VHF 23, 43, 59, 209
Waikerie 244
Warroo 109
water
   management 26, 60, 81, 157
   mining 50
   pricing 247
   rating 8, 33, 58
   recycled 7, 43, 58, 89, 248
   sources 26, 50, 60, 66, 82, 102
   temperature 33, 184
   testing 220, 243, 244
   treatment 8
   underground 171
Watergates 10, 20, 120
Weather Makers, The 6, 112, 177
weeds 85, 203

weir 98
   bifurcation 126
Brewarrina 160
Chinchilla 84
   farm 152
Glenmorgan 96
Menindee 206
Pooncarie 211
Queensland 183
St George 120
Tilpa 191
Weir 32 209
Whyenbah 128
well .... See water sources
West 2000 Plus 157
West End boat ramp 29, 144
Wetalla STP 66
white water 258
Wilcannia 199
willow trees 70, 74
Winton 11
Wivenhoe Dam 7, 8
wombat meme 106, 212, 234
woody weed 204
World Environment Day 19, 48, 55
Wyalarong Dam 8
Wycombe 109
Young Farmer of the Year 158
Young Water Professionals 229
Zuch, Andrew 38

About the Author

Steve Posselt is a civil engineer who has been active in the water industry since 1971. This has included decades of voluntary work in the Australian Water Association, a stint as director of Australia's largest water equipment manufacturing company, founding and growing a new water control gate company and, in his early days, employment in government and consulting engineering positions. He is a keen wave ski rider and he kayaks to make a difference in a world that he sees as needing a total rethink to be sustainable.

Here's what Steve has to say about this story

> It was time for an adventure. Aged 54, the time left for long endurance journeys was starting to seem finite. Brisbane to Adelaide by kayak? No-one had ever done that.
>
> You hear a lot about Australia's inland river system but what should one believe? I needed to find out for myself. As a civil engineer, passionate about water all my life, it was sobering to learn how little I know about what a river is. Much of the Darling River and its tributaries are now little more than drains.
>
> I am ashamed to be part of a society that has caused such damage. The adventure was not dangerous or extreme, but different, exciting and fun; challenging, but within the capabilities of most people, if only they would dare.

Sponsors

Without the help of the following companies, the journey through Australia's major rivers and this book would not have been possible. Everyone associated with Kayak4earth and *Cry Me a River* thanks them.

- AWA — australian water association
- Web, Multimedia & Computer Solutions — MC^2
- KSB
- Simmonds & Bristow — Helping you make good clean water!
- United Utilities
- epco australia — Specialised solutions for all your wastewater treatment needs
- Branson Brown Associates
- econnect communication
- Watergates sp
- GLOBAL — Satellite Phones & Data Communications
- BCF — BOATING • CAMPING • FISHING
- Columbia Sportswear Company
- The University Of Queensland — AUSTRALIA

Hear the truth about Australia's water

from a man who's been there

- 1 man, 1 kayak, 3 wheels
- 3 journeys, 5,000 kilometres
  - The Murray Darling
  - The Mary River
  - The East Coast

Steve Posselt

After 35 years in the water industry, Steve Posselt knows water. He has first hand experience of the state of Australia's major rivers. He has listened to mayors, engineers, farmers and bureaucrats. What he has learned about rivers and the land has come slowly, after painstaking observation on top of ten years university study. Steve always speaks openly, candidly and from the heart.

Steve can talk to your organisation

With a wealth of humour, anecdotes about outback Australians, a deep knowledge of climate change and decades of experience in managing water, Steve has developed a range of presentations suitable for school children, community groups, corporate events and technical audiences.

www.kayak4earth.com

Contact him through his web-site

Kayak4earth.com

Sustaining Australia

AT THE TIME OF GOING TO PRINT in early 2009, the Australian Government was dishing out 42 billion dollars in an attempt to save the economy. The only recognition of the dangers of climate change in that package was a rebate for insulation on houses, schools and offices.

The government was also spending around two billion dollars to buy water rights from farmers living in marginal areas to return that water to the rivers. In competition, Australia's largest private water company, Murray Irrigation Limited (MIL), was offering to buy water rights to 'protect' farmers from the government. It would allow them to keep farming if they paid MIL for the rainwater that fell on their farm, in perpetuity.

The National Farmers Federation proposed that the government's infrastructure spending be focused on making Australia the world's food bowl. Their rationale was that someone has to feed the extra 100 million people added to the world's population each year.

At the same time, the government's Carbon Pollution Reduction Scheme planned to cut emission allowances by five percent, which the Treasury estimated would increase actual emissions by almost six percent. It also proposed to compensate coal miners with billions of dollars to cover the cost of reducing their emissions and guarantee them billions more if further reductions were required.

The Liberal party responded to all this by saying that the government was endangering the economy and should only reduce greenhouse emissions and water extractions very gradually.

Environmental groups that had analysed the state of Australia's rivers focused on the environmental flow in those rivers in isolation from the causes, the landscape and the way that we would feed 20 million people.

Among all the news services that the Ebono Institute trawls every day to produce *The Generator* news, Steve Posselt was the only person to point out that land use is the key to the well-being of the rivers. Many people clamoured for the government to recognise the holistic nature of the challenges of population, water, food, energy and climate, but none of them had practical solutions for the Murray Darling.

The Institute published this book because it is the only first-hand account of the state of Australia's rivers that deals with solutions or records the opinions of people who live on the land.

Steve likes to think of his journey as an adventure. I define an adventure as containing an element of the unknown, an element of risk and an element of the taboo. Because of that combination, an adventure inevitably teaches us something. What Steve learned on his journey is something we all need to know.

The arithmetic is simple. The evaporation across inland Australia is so high, the landscape has evolved to preserve water in a variety of subtle and interlinking ways. If we interrupt those processes, we court disaster. We need to evolve our agriculture to small scale, intense and productive farms that support community and preserve the natural systems and we need to do it fast.

If publishing this book accelerates that process, we have succeeded. For our kids' sake, I hope we do.

Giovanni Ebono, publisher

Ebono Institute
Accurate analysis – incisive comment
www.ebono.org

The Ebono Institute supports authors, artists and businesses who put purpose before profit, to change the way people think, live and vote. Check out these resources from the Ebono Institute.

Guide to Saving the Planet

Giovanni Ebono's 101 tips for living more lightly on the earth, six articles and a ready reckoner that helps you calculate the footprint of various appliances and activities.
RRP: $25

Sustainable Living For Dummies

Edited by Giovanni, written by Michael Grosvenor, this book provides an overview of building a house, planting a garden and future proofing your home.
RRP: $34.95

Escape From Suburbia

The DVD that defines peak oil and then follows people who are getting off the treadmill to practise the principles of sustainability.
RRP: $35

Power Down

Richard Heinberg's analysis of the scenarios that confront us due to the depletion of oil. Written three years ago, Heinberg's detailed predictions of economic collapse, fluctuating oil prices against a background of extreme weather seem eerily pertinent today
RRP: $35